T0187216

AQA KS3

Science 2

Neil Dixon
Carol Davenport
Nick Dixon
Ian Horsewell

Approval message from AQA

This textbook has been approved by AQA for use with our qualification. This means that we have checked that it broadly covers the specification and we are satisfied with the overall quality. Full details of our approval process can be found on our website.

We approve textbooks because we know how important it is for teachers and students to have the right resources to support their teaching and learning. However, the publisher is ultimately responsible for the editorial control and quality of this book.

Please note that when teaching the *AQA KS3 Science* course, you must refer to AQA's specification as your definitive source of information. While this book has been written to match the specification, it cannot provide complete coverage of every aspect of the course.

A wide range of other useful resources can be found on the relevant subject pages of our website: www.aqa.org.uk.

HODDER
EDUCATION
AN HACHETTE UK COMPANY

Although every effort has been made to ensure that website addresses are correct at time of going to press, Hodder Education cannot be held responsible for the content of any website mentioned in this book. It is sometimes possible to find a relocated web page by typing in the address of the home page for a website in the URL window of your browser.

Hachette UK's policy is to use papers that are natural, renewable and recyclable products and made from wood grown in well-managed forests and other controlled sources. The logging and manufacturing processes are expected to conform to the environmental regulations of the country of origin.

Orders: please contact Hachette UK Distribution, Hely Hutchinson Centre, Milton Road, Didcot, Oxfordshire, OX11 7HH. Telephone: +44 (0)1235 827827. Email education@hachette.co.uk Lines are open from 9 a.m. to 5 p.m., Monday to Friday. You can also order through our website: www.hoddereducation.co.uk

© Neil Dixon, Carol Davenport, Nick Dixon and Ian Horsewell 2018

First published in 2018 by
Hodder Education,
An Hachette UK Company
Carmelite House
50 Victoria Embankment
London EC4Y 0DZ
www.hoddereducation.co.uk

Impression number   10 9 8 7 6

Year   2024

All rights reserved. Apart from any use permitted under UK copyright law, no part of this publication may be reproduced or transmitted in any form or by any means, electronic or mechanical, including photocopying and recording, or held within any information storage and retrieval system, without permission in writing from the publisher or under licence from the Copyright Licensing Agency Limited. Further details of such licences (for reprographic reproduction) may be obtained from the Copyright Licensing Agency Limited, www.cla.co.uk

Cover photo © PCN Photography / Alamy Stock Photo
Typeset by Aptara, Inc.
Printed by CPI Group (UK) Ltd, Croydon CR0 4YY

A catalogue record for this title is available from the British Library.

ISBN: 978 1 4718 9998 0

Contents

Find the answers at www.hoddereducation.co.uk/AQAKS3Science

How to get the most from this book

Transition

The blue 'Transition' pages cover the required knowledge you need from earlier study. If you are not confident with the content in the 'Core' spreads, or need more help to get started, spend time reading this material and have a go at answering the questions.

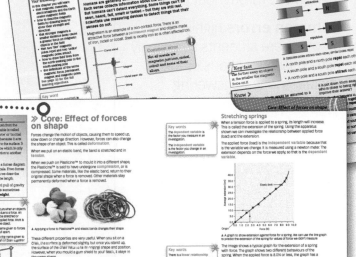

Core

The white 'Core' pages deliver the content of the AQA KS3 syllabus for each topic. These pages form the majority of the book and you should spend most of your time studying these pages.

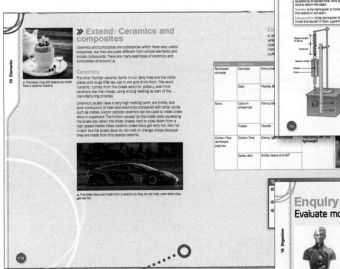

Extend

The green 'Extend' pages go into extra depth about the material you have studied on the 'Core' pages. Have a look at these pages to extend your knowledge and develop deeper understanding of the topic.

Enquiry

The yellow 'Enquiry' spreads are in-depth scenarios and activities to complete in class or at home. They will help you to present data, justify your ideas, devise useful questions and so on.

Features you will see

Key words and key facts are highlighted throughout the book.

Common errors highlight areas where students commonly have misconceptions or make mistakes when answering questions.

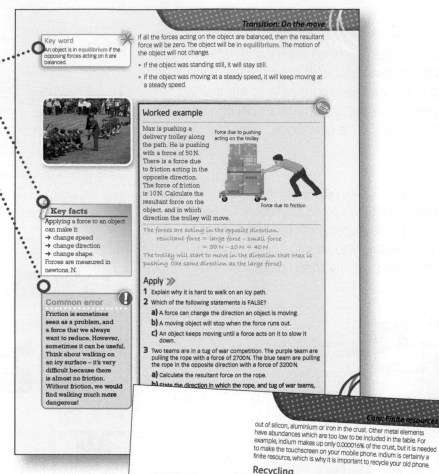

Key word
An object is in **equilibrium** if the opposing forces acting on it are balanced.

If all the forces acting on the object are balanced, then the resultant force will be zero. The object will be in **equilibrium**. The motion of the object will not change.

- If the object was standing still, it will stay still.
- If the object was moving at a steady speed, it will keep moving at a steady speed.

Worked example

Max is pushing a delivery trolley along the path. He is pushing with a force of 50 N. There is a force due to friction acting in the opposite direction. The force of friction is 10 N. Calculate the resultant force on the object, and in which direction the trolley will move.

Force due to pushing acting on the trolley

Force due to friction

The forces are acting in the opposite direction.
resultant force = large force − small force
= 50 N − 10 N = 40 N
The trolley will start to move in the direction that Max is pushing (the same direction as the large force).

Apply »

1 Explain why it is hard to walk on an icy path.
2 Which of the following statements is FALSE?
   a) A force can change the direction an object is moving.
   b) A moving object will stop when the force runs out.
   c) An object keeps moving until a force acts on it to slow it down.
3 Two teams are in a tug of war competition. The purple team are pulling the rope with a force of 2700 N. The blue team are pulling the rope in the opposite direction with a force of 3200 N.
   a) Calculate the resultant force on the rope.
   b) State the direction in which the rope, and tug of war teams,

Key facts

Applying a force to an object can make it:
→ change speed
→ change direction
→ change shape.
Forces are measured in newtons, N.

Common error

Friction is sometimes seen as a problem, and a force that we always want to reduce. However, sometimes it can be useful. Think about walking on an icy surface – it's very difficult because there is almost no friction. Without friction, we would find walking much more dangerous!

out of silicon, aluminium or iron in the crust. Other metal elements have abundances which are too low to be included in the table. For example, indium makes up only 0.000016% of the crust, but it is needed to make the touchscreen on your mobile phone. Indium is certainly a finite resource, which is why it is important to recycle your old phone.

Recycling

Key word
Recycling is when an object is processed so that the materials it is made from can be used again.

Recycling objects when they have reached the end of their usable life is important because it reduces the need to extract resources from the crust. It also saves energy, and it reduces the amount of rubbish thrown away into landfill. We recycle aluminium objects such as drinks cans because it saves energy and reduces landfill, not because we are going to run out of aluminium oxide in the crust.

▲ Aluminium cans which have been crushed, ready for recycling

Worked example

Gemma suggests that recycling magnesium is more important than recycling calcium, but Debra disagrees. Using the data in the table and pie chart, explain why Gemma could be correct.

Magnesium makes up 3% of the Earth's crust and calcium makes up 5%. Magnesium is rarer than calcium, so it is important to reduce the speed at which we extract it from the crust. Recycling magnesium is therefore more important than recycling calcium.

Know ›

1 Which is the most abundant metal element in the Earth's crust?
2 Which is the most abundant non-metal element in the Earth's crust after oxygen?
3 Give three reasons why recycling is important.

Apply »

4 Use data from the table or pie chart on the opposite page to suggest why it would be more important to recycle gold than iron.
5 Explain the benefits of recycling aluminium, despite it being so abundant in the crust.

Extend »»

6 Glass is made from sand, which is mainly silicon oxide. Using data from the table or pie chart, and your knowledge about recycling, evaluate the pros and cons of recycling glass jars and bottles.

Enquiry

7 Look at the pie chart opposite. Why is a pie chart the best way to represent this data?
8 Why are the percentages in the pie chart estimates and not accurately measured data?

Before you start answering the questions, study the **Worked example**.

The **Know questions** will test your factual recall.

The **Apply questions** ask you to explain what you have just learned or use it in unfamiliar scenarios.

The **Extend questions** are more challenging and allow you to show you really understand the topic and the ideas it covers.

Enquiry questions are short tasks that can be completed in class or at home. They help you to investigate and process data.

Your Learning objectives

On these pages we have included what are called 'Mastery Goals' from the AQA Syllabus. These are clear statements of what you need to know and how you should be able to apply the content of your course.

At the start of each major section we have created Learning objectives from the Mastery Goals. The Learning objectives are listed under the headings of Knowledge, Application and Extension.

- **Knowledge**: these objectives include important skills that need to be practised for you to become fluent; specific facts that you need to remember, and the important concepts and scientific terms.

- **Application**: these objectives are about more than remembering something or repeating a skill that you have mastered. Application is about using your knowledge and skills and being able to apply them to something new.

- **Extension**: going beyond Knowledge and Application, Extension provides more-challenging Learning Objectives where you may need to analyse how two examples are different (or similar); ask you to judge if some information is reasonable; or use what you already know to predict what might happen in a new situation.

There are copies of these Learning objectives on our website that you can print off and use as revision checklists.

You will also find the Knowledge Learning objectives at the start of each chapter, to help you see what you will be learning about.

Of course you should always check with your teacher to make sure that you are working with the most up-to-date copy of the AQA KS3 Science Syllabus and the Learning objectives that you are working towards.

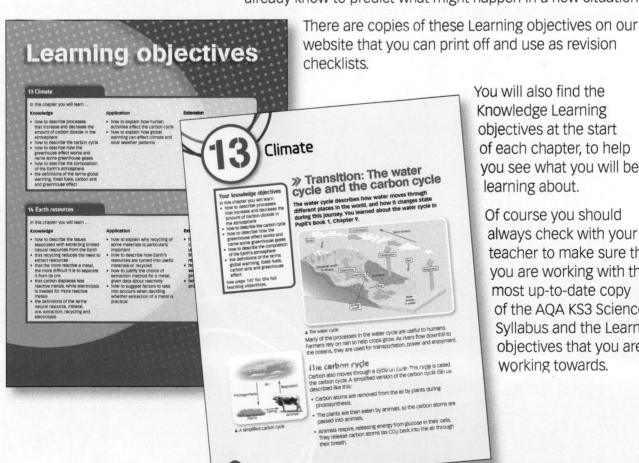

Forces

Learning objectives

1 Contact forces

In this chapter you will learn...

Knowledge

- that when the resultant force on an object is zero, it is in equilibrium and does not move, or remains at constant speed in a straight line
- how to describe how forces can change the shape of an object
- that for some objects the change in shape is proportional to the force applied

Application

- how to explain whether an object is in equilibrium
- how to describe factors that affect the size of frictional and drag forces
- how to describe how materials behave as they are stretched or squashed
- how to describe what happens to the length of a spring when the force on it changes
- how to use force and extension data to compare the behaviour of different materials in deformation using the idea of proportionality

Extension

- how to evaluate how well sports or vehicle technology reduces frictional or drag forces
- how to describe the effects of drag and other forces on falling or accelerating objects as they move
- how to explain how turning forces are used in levers

2 Pressure

In this chapter you will learn...

Knowledge

- how to describe how pressure acts in a fluid
- the effect of different stresses on a solid object when it is scratched or broken
- the definition of the term atmospheric pressure

Application

- how to explain observations of fluids in terms of unequal pressure using diagrams
- how to explain why objects either sink or float using weight and upthrust
- how to explain observations in which the effects of forces are different because of differences in the area over which they apply
- how to use the formula: fluid pressure, or stress on a surface = force (N)/area (m²)

Extension

- how to use the idea of pressure changing with depth to explain underwater effects
- how to carry out calculations involving pressure, force and area in hydraulics, where the effects of applied forces are increased
- how to use the idea of stress to deduce potential damage to one solid object by another

1 Contact forces

≫ Transition: On the move

In order to move, a rollerblader has to apply a force to the road. The rollerblader can move quickly and easily on a smooth tarmac road. However, when the road becomes rough, she cannot skate as fast. The friction between the skates and the road surface affects her speed.

Your knowledge objectives
In this chapter you will learn:
- that when the resultant force on an object is zero, it is in equilibrium and does not move, or remains at constant speed in a straight line
- how to describe how forces can change the shape of an object
- that for some objects the change in shape is proportional to the force applied

See page 7 for the full learning objectives.

Key words

Friction is the force between two surfaces that are sliding over each other, or that are trying to slide over each other. Friction always opposes the direction of motion.

A **resultant force** is a single force that could replace all of the forces acting on an object and still have the same effect on the object.

▲ A smoother surface has less friction and makes skating easier

What can forces do?

We often think of forces as 'pushes' or 'pulls'. When we push or pull objects then we change how they move. They might speed up or slow down or change direction. Whenever we see the motion of an object change, we know that there must be a force. One or more forces acting on an object can also change its shape.

A rolling ball comes to a stop because the force of friction slows it down.

The bigger the force acting on an object, the more it will speed up or slow down.

Common error

Some people think that a force has to be acting on an object if it is moving at a steady speed. This isn't true. When an ice hockey puck is hit, it will travel at a steady speed. There is almost no friction on the ice, so there are no forces acting in the direction that the puck is moving.

Combining forces

There is often more than one applied force acting on an object. To work out how the object will move we have to work out the **resultant force** acting on it.

The table shows how you can calculate the resultant force for forces that are all acting in a straight line.

Direction of the applied forces	Resultant force
Forces in the **same** direction	Add the forces
Forces acting in **opposite** directions	Largest force – smallest force

Key word

An object is in **equilibrium** if the opposing forces acting on it are balanced.

If all the forces acting on the object are balanced, then the resultant force will be zero. The object will be in **equilibrium**. The motion of the object will not change.

• If the object was standing still, it will stay still.

• If the object was moving at a steady speed, it will keep moving at a steady speed.

Key facts

Applying a force to an object can make it:

→ change speed
→ change direction
→ change shape.

Forces are measured in newtons, N.

Common error

Friction is sometimes seen as a problem, and a force that we always want to reduce. However, sometimes it can be useful. Think about walking on an icy surface – it's very difficult because there is almost no friction. Without friction, we would find walking much more dangerous!

Worked example

Max is pushing a delivery trolley along the path. He is pushing with a force of 50 N. There is a force due to friction acting in the opposite direction. The force of friction is 10 N. Calculate the resultant force on the object, and in which direction the trolley will move.

Force due to pushing acting on the trolley

Force due to friction

The forces are acting in the opposite direction.

resultant force = large force – small force
= 50 N – 10 N = 40 N

The trolley will start to move in the direction that Max is pushing (the same direction as the large force).

Apply »

1 Explain why it is hard to walk on an icy path.

2 Which of the following statements is FALSE?

a) A force can change the direction an object is moving.

b) A moving object will stop when the force runs out.

c) An object keeps moving until a force acts on it to slow it down.

3 Two teams are in a tug of war competition. The purple team are pulling the rope with a force of 2700 N. The blue team are pulling the rope in the opposite direction with a force of 3200 N.

a) Calculate the resultant force on the rope.

b) State the direction in which the rope, and tug of war teams, will move.

★ **Pupil's Book 1, Chapter 2 discussed contact and non-contact forces.**

Key word

A **contact** force is one which acts only when objects are in contact with each other. Once the contact is removed the force is no longer applied.

★ **'Forces spectacles' aren't a real thing! They are an idea which helps us to visualise the forces that act on an object.**

▲ One of the authors wearing 'forces spectacles' to help her see the forces acting on objects

Common error ❗

Some people think that the table 'just gets in the way' and doesn't apply a force to the book. However, if there wasn't an equal but opposite normal contact force balancing the force due to gravity, the table would break and the book would continue to move downwards.

≫ Core: Force diagrams

The forces acting on an object can be **contact** forces, or non-contact forces. We can see some contact forces being applied, especially if a person is applying the force by pushing or pulling. However, for most forces acting on objects, we can only see their effects.

To understand the motion of objects when forces are applied, we draw force diagrams. It's like we have put on 'forces spectacles' to help us see the invisible forces acting.

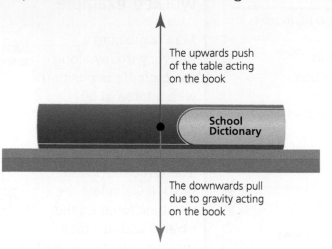

The upwards push of the table acting on the book

School Dictionary

The downwards pull due to gravity acting on the book

▲ A force diagram to show the forces acting on a book on a table

When you look at the book lying on a table in the left-hand picture above you can't see any forces acting on it. Its motion isn't changing so it is in equilibrium.

The right-hand picture shows the same book when looked at through forces spectacles. Forces arrows represent the size of each force on the book. Now we can see that there are two forces acting on the book. There is a downwards pull due to gravity, and a balancing upward push from the table.

We can use simple rules to label the forces acting on an object.

1 Name the forces acting on the object.

2 Identify the direction of the forces acting on the object.

3 Add force arrows:

   a) The length of the arrow shows the size of the force.

   b) The direction of the arrow shows the direction of the force.

   c) The label describes the action of the force.

Using these rules means that we know exactly what the forces acting on the object are, and what effect they are having on the object. Later on in your physics studies you will also be able to label the arrows with the size of the force in newtons.

Common error

We know that gravity pulls down on all of the book. It's tempting to put lots of force arrows over the book to represent this. However, to simplify things we can treat forces as if they act at one point. This is shown by the black circle in the photo opposite.

Worked example

Draw and label the forces acting on an apple hanging on a tree as shown in the picture on the right.

The apple isn't moving, so we know that the forces acting on it are in equilibrium. There is a force pulling downwards due to gravity acting on the apple. There is a force pulling upwards due to the tension in the branch. The two arrows are the same size.

Upwards pull on apple due to tension in branch

Downwards pull due to gravity acting on the apple

Know >

1 Copy and complete the table with examples of contact and non-contact forces. Two have been filled in already.

Contact forces	Non-contact forces
Friction	Gravity

2 The forces on an object are in equilibrium. State what will happen to the motion of the object.

Apply >>

3 Copy each of the following diagrams, and draw and label the forces acting on the object.

a) Ship stationary at sea

b) Ice hockey puck moving at a steady speed

c) Rocket launch

4 Describe what would happen to the apple in the worked example if the force due to tension in the branch was 0.5 N and the force due to gravity was 1 N.

Extend >>>

5 The ship shown in the diagram above starts its engines and starts to accelerate forwards. Draw and label the forces acting on the ship now.

6 Kai is out cycling. He applies the brakes.

a) Draw and label the forces acting on the bike as it slows down.

b) State and explain what forces are acting on the bike once it is stationary.

Key facts

Key facts

The upward push from the surface of the table is called the 'reaction force' or 'normal contact force', because it acts at right angles to the surface. It is a contact force which is only present in reaction to another force acting.

The arrows on a forces diagram are drawn to scale. If two forces are equal, then we draw the arrows the same length.

The downward pull of gravity on an object is sometimes labelled as **weight**.

Key words

Deformation occurs when an object's shape is changed due to a force. An elastic object can be stretched or squashed by an applied force. Work is done to deform the object.

Tension is the name given to forces that extend or pull apart.

Compression is the name given to forces that squash or push together.

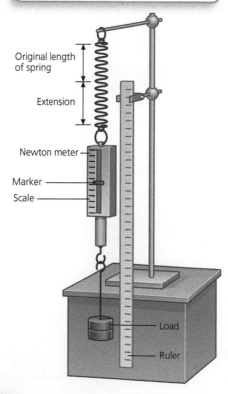

Original length of spring

Extension

Newton meter

Marker

Scale

Load

Ruler

» Core: Effect of forces on shape

Forces change the motion of objects, causing them to speed up, slow down or change direction. However, forces can also change the shape of an object. This is called **deformation**.

When we pull on an elastic band, the band is stretched and in **tension**.

When we push on Plasticine™ to mould it into a different shape, the Plasticine™ is said to have undergone **compression**, or is compressed. Some materials, like the elastic band, return to their original shape when a force is removed. Other materials stay permanently deformed when a force is removed.

▲ Applying a force to Plasticine™ and elastic bands changes their shape

These different properties are very useful. When you sit on a chair, the surface is deformed slightly, but once you stand up, the surface of the chair returns to its original shape and position. However, when you mould a gum shield to your teeth, it stays in the same shape.

▲ Responding differently to deformation. (a) Chairs return to their original shape, (b) after being moulded, gum shields keep their new shape

◄ Using this apparatus we can investigate how forces affect the length of a spring

Stretching springs

Key words

The **dependent variable** is the factor you measure in an investigation.

The **independent variable** is the factor you change in an investigation.

When a tension force is applied to a spring, its length will increase. This is called the extension of the spring. Using the apparatus shown we can investigate the relationship between applied force (load) and the extension.

The applied force (load) is the **independent variable** because that is the variable we change. It is measured using a newton meter. The extension depends on the force we apply, so that is the **dependent variable**.

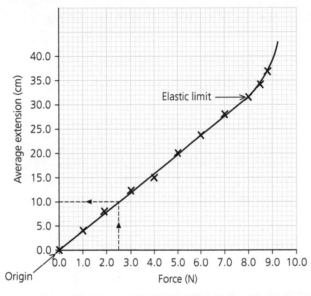

▲ A graph to show extension against force for a spring. We can use the line graph to predict the extension of the spring for values of force we didn't measure

Key words

There is a **linear relationship** between two variables if there is a straight line when they are plotted on a graph. If the straight line goes through the origin, then they are described as **directly proportional** to each other.

The image shows a typical graph for the extension of a spring with force. The graph shows two different behaviours of the spring. When the applied force is 8.0N or less, the graph has a **linear relationship** and shows that the extension is **directly proportional** to the applied force. If we remove the force the spring will return to its original length.

However, when the applied force increases above 8.0N, the graph starts to curve. Increasing the force by the same amount causes the extension to be much greater. This is called the elastic limit, and if we remove the load, the spring will stay deformed.

▲ These two springs were the same length. One has been stretched beyond its elastic limit and is permanently deformed

Worked example

Esta measured the extension against force for a spring. The graph on the right shows her results.

1 Esta used a maximum force of 3N so that the spring wasn't permanently deformed. Suggest how Esta could improve her data.

2 Use the graph to:

   a) work out the extension of the spring if a force of 1.5N was applied to it

   b) work out the force applied if the extension of the spring is 15cm.

1 Esta has taken four measurements, with only three measurements with a force applied to the spring. She should also have taken measurements with applied forces of 0.5N, 1.5N and 2.5N, which would have given her six data points.

2 a) Draw a vertical line from the x-axis for force at 1.5N until it meets the line of best fit. Then draw a horizontal line from this point and read off the value of extension on the y-axis. This is shown by the green line on the graph on the right. When force = 1.5N, extension = 10.5cm.

   b) Draw a horizontal line from the y-axis for an extension of 15cm until it meets the line of best fit. Then draw a line vertically down from this point and read off the value of force on the x-axis. This is shown by the purple line on the graph. When extension = 15cm, force = 2.15N.

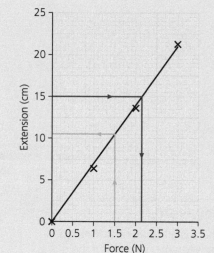

Know >

1 A spring has a compression force applied. What does this mean?

2 Give an example of when an object returning to its original shape after a force is removed is a useful property.

3 Give another example of a material that permanently deforms after a force has been applied.

Apply ≫

4 Esta uses a different spring and obtains the mean values for extension shown in the table.

Applied force (N)	Extension (cm)
0.0	0.0
0.5	0.7
1.0	1.5
1.5	2.6
2.0	3.4
2.5	4.4
3.0	5.3

a) Esta took three readings for each value of applied force. Describe how to calculate the mean value for extension.

b) Plot a graph of extension (on the *y*-axis) against applied force (on the *x*-axis) using the data in the table.

c) Draw a line of best fit on your graph.

d) Use your graph to describe how the extension of the spring varies with applied force.

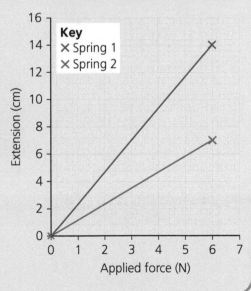

Extend ≫≫

5 Explain why the car bonnet shown in the picture has not returned to its original shape after a car crash.

Enquiry ≫≫

6 Springs may be made out of different materials, which stretch different amounts. The diagram below shows the extension–force graph for two different springs.

Use the graph to calculate the extension of each spring when:

a) a force of 3 N is applied

b) a force of 6 N is applied.

7 Use the graph in question 6 to decide which of the two springs is the 'stretchiest'. Explain your answer.

Key
× Spring 1
× Spring 2

(Graph: Extension (cm) on y-axis, 0 to 16; Applied force (N) on x-axis, 0 to 7)

Key words

Drag occurs when two surfaces move over each other and one of the surfaces is a fluid. It is a frictional force which opposes motion.

A **fluid** is a substance that has no fixed shape. Both gases and liquids are fluids.

» Extend: Drag forces

Wading through water is hard work. The **drag** from the water opposes your motion and slows you down. Air resistance is another example of drag. Drag is the name given to friction when one of the surfaces involved is a **fluid**. In physics, a fluid can be a liquid or a gas, because although they look different, they can both flow.

Designing faster swimsuits

In many sports there is a competitive advantage to reducing the frictional and drag forces involved. One sport in which this has happened is swimming.

Hi-tech swim suits to be thrown out of the pool

The hi-tech swim suits that made such a splash at the Beijing Olympics in 2008 will be banned from 1st January 2010. Swimmers using these suits have broken all but two of the world records.

The designers of the stylish full-body suits have modelled them on shark skin. The suits reduce the swimmer's drag in the water, letting them swim much faster.

Extract from news article, July 2009

▲ Hi-tech LZR swimsuits unveiled in February 2008

Tasks

Water surface

1. Draw a diagram showing the forces acting on a swimmer racing through the water.
2. Discuss possible views for and against a ban on hi-tech swimsuits in competitive swimming.

Key word

A **lever** is a simple mechanism that can be used to allow a small force to have a larger effect. It pivots about a point, known as a fulcrum.

▲ Using a spoon as a lever to open a tin of syrup

▲ All these tools use turning forces. Can you describe how they work?

▲ This maths 'see-saw' can be used to investigate turning effects

Levers

Forces can have a turning effect. A **lever** is one example of when this can happen. Levers are simple mechanisms that allow an applied force to have a greater effect by using a fixed point that the object pivots round.

As the handle of the spoon in the picture is pushed down, it cannot move straight down, but instead pivots clockwise about the fulcrum. This means that there is an anticlockwise turning force at the bowl of the spoon, which allows us to prise open the tin. If we measure the size of the two forces, then the initial applied force will be smaller than the force acting at the tin lid.

Investigating turning effects

Using the apparatus shown below left, Jiao investigates how the turning force and distance from the pivot point affect the motion of the maths see-saw. Jiao adds different numbers of hangers to one peg on each side so that the see-saw is balanced. Her results are shown below.

Left-hand side of see-saw		Right-hand side of see-saw	
Number of hangers	Distance from pivot (cm)	Number of hangers	Distance from pivot (cm)
1	1	1	1
1	2	2	1
2	4	1	8
2	10	5	4
3	2	1	6
3	2	2	3

Tasks

5. Jiao puts one hanger at 6 cm from the pivot on the left-hand side of the see-saw. Write down three different ways Jiao could put hangers on the right-hand side to balance the see-saw.
6. Describe the relationship between the turning forces on the maths see-saw.
7. Jiao puts one hanger at 10 cm from the pivot on the left-hand side of the see-saw. She then puts one hanger at 3 cm and one hanger at 6 cm from the pivot on the right-hand side of the see-saw. State and explain whether the see-saw is balanced.

Enquiry:
Safe landings from space

Upwards force due to air resistance acting on the parachute and the craft

Downwards force due to gravity acting on the craft

▲ The size of the forces acting on the landing craft as the parachute opens. The craft travels at a steady speed, towards the ground

▲ Using a parachute to land a Soyuz spacecraft safely

▶ The crash site where Schiaparelli landed. Photograph taken from orbit around Mars

Once spacecraft have travelled to other planets, it's important that they can land safely. This involves finding a good landing point. It's also important to make sure that the landing craft is travelling slowly enough to touch down, and not crash.

As any object moves through the atmosphere, air resistance provides a force in the opposite direction to the motion.

On Earth, parachutes are a good way to slow down a landing craft once it has entered the atmosphere. The diagram on the left shows the forces that are acting on a spacecraft once the parachute has opened.

On the Moon, there is no atmosphere and therefore no air resistance. Parachutes can't be used to help land spacecraft. They have to use rockets to fire downwards. This provides an upward force to balance the downward force due to gravity.

Mars has a much thinner atmosphere than the Earth. The air pressure on the surface is less than 1% of that of the Earth. This makes it much harder to land spacecraft on Mars. The atmosphere means that the spacecraft will heat up as it travels towards the surface. However, the low air resistance means that parachutes have to be very large to provide sufficient balancing force.

In October 2016, the European Space Agency ExoMars mission successfully reached Mars and entered orbit around the planet. However, the Schiaparelli lander craft crashed as it tried to land on the surface of the planet.

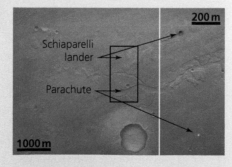

200 m

Schiaparelli lander

Parachute

1000 m

★ **Useful websites about space are www.nasa.gov and www.esa.int**

❶ Use the internet (or other sources) to find out more about other exploration missions to Mars, and how the landing vehicle reached the surface. Two interesting missions you could find out about are:
- Beagle 2 (launched 2 June 2003)
- Curiosity (launched 26 November 2011)

Factors affecting drag forces

Air resistance and drag slow objects down. Cupcake cases can be used to investigate the factors that affect drag forces.

When a cupcake case is dropped, there are two forces acting on it: the downward force due to gravity (weight) and the upward force due to drag. If these two forces are equal, then the cupcake case will fall at a steady speed. This is known as terminal velocity.

② Draw a forces diagram to show the forces acting on a cupcake case as it falls at its terminal velocity.

▲ Different sized cake cases to test the effect of area and mass on drag force

Abeo is investigating how mass affects the terminal velocity of cupcake cases. He drops a cupcake case and uses a motion sensor and data logger to measure the speed of the cupcake case after it has fallen 2 m. He repeats this measurement five times because using the data logger means he can export his data into a spreadsheet and process it quickly.

To increase the mass of the cupcake case, Abeo stacks two cupcake cases together and drops them. He repeats this until he has 10 cupcake cases in a stack.

③ Suggest one other factor that could affect the terminal velocity of a cupcake case.

④ What variables will Abeo need to keep constant during his investigation?

▲ Apparatus to measure how fast cupcake cases fall

The graph that Abeo obtained from his investigation is on the left.

⑤ Use the graph to describe the relationship between number of cupcake cases and speed.

⑥ Explain the shape of the graph using ideas about forces.

⑦ Suggest how the investigation could be done without the use of a data logger. What measurements would need to be taken?

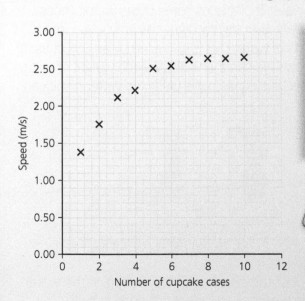

Key fact

→ On the Moon, all objects will fall at the same rate because there is no air resistance.

② Pressure

See page 7 for the full learning objectives.

See page 7 for the full learning objectives.

Your knowledge objectives

In this chapter you will learn:

- how to describe how pressure acts in a fluid
- the effect of different stresses on a solid object when it is scratched or broken
- the definition of the term atmospheric pressure

See page 7 for the full learning objectives.

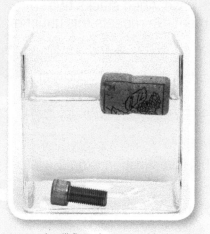

▲ A cork will float, but a steel bolt will sink

Key word

Upthrust is the upward force that a liquid or gas exerts on an object floating in it.

Common error

Children often think that only light objects will float, and all heavy objects will sink. But huge ships float, and they are very heavy.

» Transition: Floating and sinking

Put some objects in water, and they will float on the surface. Put other objects in water and they will sink. To understand why this happens, we need to think about the forces acting on objects in water.

The cork in the photo is floating at the surface of the water. It is stationary, so the forces on it must be balanced. There will be a downwards force due to the weight of the cork. There must therefore be a balancing upwards force acting on the cork. This force is called **upthrust**. We can draw a force diagram to show the forces on the cork.

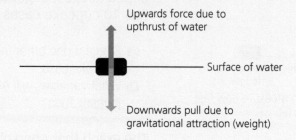

Upwards force due to upthrust of water

Surface of water

Downwards pull due to gravitational attraction (weight)

When an object is put it water, it displaces some of the water. That's why the water level in a bath gets slightly higher when you get into it. We can measure the mass of water that is displaced by an object using the apparatus shown below. We can then calculate the weight of water displaced. The upthrust force is equal to the weight of water displaced.

► The Eureka can apparatus used to measure the amount of water displaced by a floating block

By comparing the weight of the object with the weight of the water displaced we can decide whether an object will float or sink.

- Weight of the water displaced **equals** weight of the object – object will float.

- Weight of water displaced is **less than** the weight of the object – object will sink.

Worked example

A student is investigating floating and sinking. She has a ball of Plasticine™ that weighs 1 N and a Eureka can apparatus. When she puts the ball into the water, it sinks to the bottom of the can. It displaces water with a weight of 0.6 N.

She then makes a boat shape using the Plasticine™. She refills the Eureka can and floats the boat in the water. The Plasticine™ boat displaces water with a weight of 1 N.

Explain why the boat floats but the ball doesn't.

The Plasticine™ boat displaces water equal to its own weight (1 N). The upthrust is equal to the weight, so the boat floats in the water. The Plasticine™ ball displaces water with a weight of 0.6 N. The upthrust is less than the force due to gravity pulling the ball down, so it sinks.

Apply »

1 A canoe displaces water with a weight of 100 N when it floats.

   a) What is the weight of the canoe?

   b) Calculate the mass of the canoe.

2 Water walkers are giant inflatable balls that let people walk on water.

   a) Explain why water walkers float.

   b) Predict what would happen to the position of the water walker if a heavier child was inside.

▲ Walking on water

3 Pumice and obsidian are both types of stone. When two stones of similar size are put in water, the pumice floats, but obsidian doesn't. Explain why the two stones behave differently in water.

▲ Although steel bars sink, a ferry made of steel will float because it can displace water equal to its own weight

» Core: Forces on a surface

Pushing down

Whenever an object rests on a solid surface there is a force due to gravity acting downwards. The effect that this force has on the surface depends on the area the force is acting over. This is known as **stress**.

The stress will be decreased if the force is acting over a larger area, like walking in snowshoes. The stress will be increased if the force is acting over a smaller area, for example when using a cooking knife.

Key word

The **stress** on a surface is the ratio of applied force to surface area. The units of stress are N/m².

▲ Snowshoes stop you sinking into the snow, by reducing the stress at the surface

Calculating stress

We can calculate the stress acting on a surface if we know the applied force and the area that the force is acting over. Often the force will be the weight of the object resting on the surface.

$$\text{stress (N/m}^2) = \frac{\text{force (N)}}{\text{area (m}^2)}$$

If the stress on a surface is too large, the surface can be damaged and permanently deformed. Very large stresses can lead to the surface breaking.

Common error

You will sometimes see stress on a surface described as pressure. However, pressure is a term that should be used with fluids, not solids.

You may have been told to 'keep all four legs of your chair on the floor' by a teacher or parent. One reason for this is that you are more likely to fall off the chair if it's balanced on one or two legs. However, it's also because when all four legs are on the floor the force is distributed equally over them, reducing the stress. The stress on the floor doubles if you rock on two chair legs, and quadruples if you swing on one leg. This stress can be large enough to damage the surface of the floor.

Worked example

A large box is placed on a table. The box has a mass of 9 kg. The box covers an area of 0.21 m². What is the stress on the table surface due to the box?

The force is equal to the weight of the box. Remember:

weight (N) = mass (kg) × gravitational field strength (N/kg)

So the force = 90 N

$$\text{stress (N/m}^2) = \frac{\text{force (N)}}{\text{area (m}^2)}$$

$$\text{stress} = \frac{90\,N}{0.21\,(m^2)} = 428.57\,N/m^2$$

Know>

1 What is the unit of stress?

2 A brick has a mass of 2.8 kg. Calculate the weight of the brick.

3 The base of the brick is 102.5 mm × 215 mm. Calculate the area of the brick.

Apply>>

4 The force of a drawing pin on your thumb and on the surface is the same. But the drawing pin goes into the surface, not your thumb. Explain why.

5 A ballet dancer is balanced on one foot in pointe shoes. The area of her shoe in contact with the floor is 10 cm². She has a mass of 50 kg. Calculate the stress on the floor while she is standing on pointe.

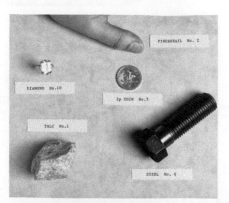

◄ Different common items and their Mohs hardness number. The higher the number the harder it is to scratch the material

Extend》》》

6 An African elephant has a mass of 6000 kg. The diameter of each foot is approximately 0.4 m. Calculate the stress on the surface underneath each foot.

Enquiry》》》

Different materials will respond differently to the same level of stress at the surface. Some materials are easily scratched, but others are very resistant to scratches. Materials scientists call this property hardness. The Mohs hardness scale is used to measure how resistant different materials are to being scratched. It was originally developed to identify different rocks and minerals. Diamond is the hardest material on the scale (number 10) and talc is the softest (number 1).

Object	Mohs hardness scale
Bathroom tile	6.5
Copper coin	3
Diamond	10
Granite worktop	7
Marble worktop	3
Steel knife blade	6
Window glass	5.5

If two objects are scratched together, then the object with the higher Mohs hardness number will scratch the other object. If the two objects have the same hardness number, they will both scratch each other.

7 Use the information in the table to determine which of the two objects will be scratched:

a) copper coin and steel knife blade

b) diamond and window glass

c) bathroom tile and steel knife blade

d) copper coin and marble worktop.

8 Suggest why granite is more popular as a kitchen worktop than marble.

9 You find an unknown mineral rock. Describe how you could use objects from the table to measure the Mohs hardness of the rock.

» Core: Pressure in a fluid

Under pressure

★ **1000 Pa = 1 kPa**

The pressure in a swimming pool depends on the weight of water above you. The weight applies a force which acts over the area of your body. The deeper you go in a pool, the greater the weight of water above you, and the higher the **pressure**.

If you climb to the top of a mountain, then the air pressure is lowered. Although you don't notice it, the column of air above you exerts a force over your surface area. You feel this as pressure. The less air there is above you, the lower the pressure. **Atmospheric pressure** at sea level is about 100 kPa. This is often called 1 atmosphere (1 atm). Air pressure at the top of Everest is about 30 kPa.

Key words

Pressure is the ratio of force to surface area for a fluid. The unit is N/m^2 or Pascals (Pa).

Atmospheric pressure is the pressure caused by the weight of the air above a surface.

▲ Diving underwater can hurt your ears because of the increased pressure at the bottom of the pool

▲ Demonstrating that water pressure is greater as water depth increases

Explaining pressure

Pressure in a fluid is caused in the same way whether the fluid is a liquid or a gas. Imagine standing in a ball pool. If someone throws a ball at you, you will feel the force of it as it hits. If lots of balls hit you all at once, then you would feel the force in lots of different places, and probably wouldn't be able to tell where you were being hit.

★ **You can read more about the particle theory of matter in Pupil's Book 1, Chapter 9.**

In the same way, pressure is caused by the particles in the fluid moving at random and bouncing off a surface. Each time they bounce off they apply a very small force to the surface. Although this is a tiny force and you don't feel it, over the whole surface it averages out to a pressure you can feel.

To calculate the pressure in a fluid we use the equation:

$$\text{pressure (Pa)} = \frac{\text{force (N)}}{\text{surface area (m}^2\text{)}}$$

Pushing back

You can't feel air pressure because your insides 'push back' with the same pressure (or force per unit area). However, if you put marshmallows in a bottle and then suck out some of the air, you can see the marshmallows expand. They are soft and contain pockets of gas. The air pressure in the bottle is reduced, but the gas pressure in the marshmallows stays the same. This causes the marshmallows to expand.

▲ The pressure in a fluid acts at all points of a surface, pushing inwards

▲ The marshmallows expand when the surrounding pressure is reduced

Worked example

An Olympic-sized swimming pool is 50 m long and 25 m wide. It contains 2 500 000 kg water. Calculate the water pressure at the base of the swimming pool.

$$\text{pressure (Pa)} = \frac{\text{force (N)}}{\text{surface area (m}^2)}$$

The force of the water is equal to the weight of the water.

$$\text{weight} = 2\,500\,000\,\text{kg} \times 10\,\text{N/kg} = 25\,000\,000\,\text{N}$$

$$\text{surface area} = \text{length} \times \text{width} = 50\,\text{m} \times 25\,\text{m} = 1250\,\text{m}^2$$

$$\text{pressure (Pa)} = \frac{25\,000\,000\,\text{N}}{1250\,\text{m}^2} = 20\,000\,\text{Pa}$$

Know>

1 A 250 ml beaker contains 0.25 kg of water. Calculate the weight of the water.

2 What two variables do you need to know to calculate pressure?

3 Describe the difference between liquids and gases in the particle model.

Apply>>

4 A paving stone has an area of 1 m². The mass of the air column above the stone is 10 000 kg. Calculate the air pressure on the paving stone.

5 Look at the marshmallows in the photo on page 21. Predict what will happen when the air is 'let back into' the bottle. Explain your answer.

Extend »»

A classic demonstration of air pressure is the 'collapsing can' demonstration. A small amount of water is put into a drinks can. The water is then heated until it boils. The can is quickly turned upside down and plunged into cold water. The can immediately collapses.

▲ Collapsing can demonstration

6 Explain why the can is crushed by the air. Use ideas about particles and air pressure in your answer.

Enquiry »»»

Researchers use weather balloons to measure properties of the atmosphere. This can include temperature, levels of different gases and wind speeds.

The balloon is filled with hydrogen or helium. When it is released there is an upthrust acting on the balloon, and it rises upwards in the air.

7 Draw a forces diagram to show the forces acting on the balloon.

8 Use your forces diagram to explain why the balloon moves upwards.

As the balloon rises into the atmosphere it expands, and its diameter increases. Eventually it bursts, and the instrument packet falls back down to earth. The packet has a parachute attached to prevent damage as it lands.

9 Explain why the balloon expands and bursts.

10 How does the parachute help prevent damage to the packet?

11 Another way to measure the atmosphere would be to use a plane or a helicopter to collect data. Discuss some of the benefits and disadvantages to the researchers of measuring the atmosphere using a weather balloon.

▲ A researcher in Antarctica releases a weather balloon which carries an instrument package high into the sky to measure the atmosphere

» Extend: Hydraulics

One property of liquids is that they hardly change their volume when they are compressed. This means that if you try to compress a liquid it won't be squashed and it will transmit the pressure through the liquid. This property is used in hydraulic systems such as jacks and car brakes.

▲ A hydraulic jack enables a person to easily lift up a car

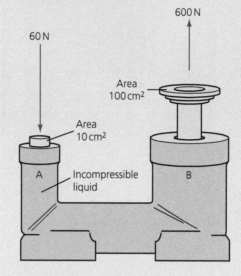

▲ A simplified diagram showing how a hydraulic jack works

Key word

An **incompressible liquid** is an ideal liquid which doesn't change volume when a force is applied. Most real liquids, such as water, will be squashed slightly if a force is applied.

The diagram above shows a force of 60 N being applied to the handle of a hydraulic jack at A. In this simplified model, the jack is filled with an **incompressible liquid**. The force causes an extra pressure in the liquid.

$$\text{pressure} = \frac{\text{force (at A)}}{\text{area of A}} = \frac{60\,\text{N}}{10\,\text{cm}^2} = 6\,\text{N/cm}^2$$

The pressure at A is 6 N/cm². This pressure is passed through the liquid. Therefore, at B, where the jack is underneath the car, the pressure is also 6 N/cm². There will be an upwards force on the surface at B.

$$\text{force (at B)} = \text{pressure} \times \text{area of B} = 6\,\text{N/cm}^2 \times 100\,\text{cm}^2 = 600\,\text{N}$$

Because the surface area at B is greater than the surface area at A, the hydraulic jack acts as a force multiplier. Using a small applied force, we can lift heavy objects.

Tasks

1. A force of 100 N is applied to the jack shown in the diagram. Calculate the weight of the object that the jack can lift at B.
2. Calculate the force that needs to be applied to the jack to lift a weight of 5000 N.

Underwater effects

▲ The structure of the dam ensures that the force of the water doesn't cause it to collapse

The further you go under water, the greater the water pressure gets. The force applied to your body surface increases because there is a greater weight of water above.

Dams are built so that the base is thicker than the top. This is because the force at the base of the wall is greater due to the increased water pressure.

Task

③ The lake behind the Hoover Dam is 220 m deep. The water pressure at the bottom of the dam is 2 000 000 Pa. Calculate the force of the water acting on an area of 1 m² at the bottom of the dam.

▲ Lantern fish live deep in the ocean and are adapted to cope with the high pressures there

Lantern fish are deep-sea fish. They can live at depths of more than 500 m. There is an urban myth that when deep-sea fish are brought to the surface they explode. This does not appear to be the case.

Fish that live closer to the surface often have a swim bladder. By controlling the amount of gas in the swim bladder fish can move up and down in the water. The external water pressure is balanced by the internal pressure of the gas in the swim bladder and the rest of the fish. If the water pressure decreases the gas will expand, and vice versa.

Deep-sea fish often don't have a swim bladder so that the changes in pressure don't affect them. That means that they don't explode when brought to the surface. In fact, evolutionary biologist Craig McClain says that as long as deep-sea creatures are kept cool, they can survive well at lower pressures than they are used to.

▲ The increased pressure at the bottom of the sea has caused the gas pockets in the expanded polystyrene to collapse

Task

④ Research other ways in which fish are adapted to cope with the increased pressure at the bottom of the sea.

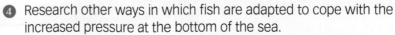

Enquiry:
Under the sea

★ **Boaty will be based on the Polar exploration ship *RRS Sir David Attenborough*. One name suggested for the ship was 'Boaty McBoatface'. The sub-sea vehicle is named after this suggestion.**

'Boaty' is a remote-controlled sub-sea vehicle being built to explore under the seas around the Arctic and Antarctic. It will be used by scientists to learn more about these difficult to reach areas.

▲ The sub-sea research vehicle Boaty McBoatface

When a sub-sea vehicle goes under the water, it experiences an upthrust force. The upthrust is equal to the weight of water that the vehicle displaces.

The diagram below shows the forces that act on the object as it floats in the water.

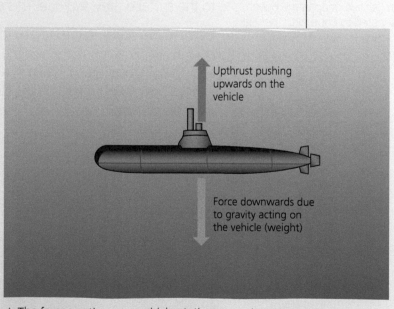

Water surface

Upthrust pushing upwards on the vehicle

Force downwards due to gravity acting on the vehicle (weight)

▲ The forces acting on a vehicle stationary under water

★ **The upthrust force is sometimes referred to as buoyancy.**

The upthrust also depends on density and temperature of the water that the object is in. A higher density will give a greater upthrust because the water displaced has a greater weight. A higher temperature will give a smaller upthrust because the density of the water is lower.

❶ A sub-sea vehicle has a weight of 6000 N when it is stationary. State the upthrust force on it when under the surface of the water.

❷ Draw force diagrams which show the forces acting on the vehicle when it is:
  a) moving forwards at a steady speed
  b) moving forwards and speeding up.

❸ Sea water has an average density of $1.025\,g/cm^3$. Pure water has a density of $1.000\,g/cm^3$. Explain why the upthrust on Boaty is greater when floating under the surface of the sea.

Planning an investigation into upthrust

When an object is attached to a newton meter and then submerged in water it will appear to have a lower weight than when in air.

The difference between the weight in air and weight in water is equal to the upthrust on the object.

The photograph below shows a simple apparatus that can be used to investigate upthrust. A small lump of Plasticine™ is suspended from a newton meter. The reading on the newton meter is taken when the Plasticine™ is hanging in air. The Plasticine™ is then fully submerged in water, and the new reading on the newton meter is taken.

The difference between these two readings gives the value of upthrust.

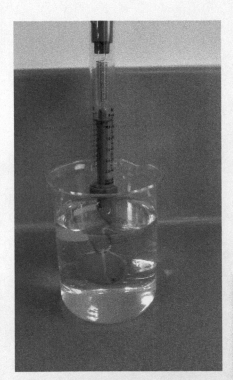

▲ Using a newton meter to measure the apparent weight loss of a ball of Plasticine™

❹ Plan an investigation into how the density of the water affects the upthrust on an object. You should address the following questions:
  a) What are the dependent and independent variables?
  b) What are the control variables?
  c) How will you alter the density of the water?
  d) What measurements will you need to take? How many?
  e) What method will you use?

Electromagnetism

Learning objectives

3 Magnetism

In this chapter you will learn...

Knowledge

- that magnetic materials, electromagnets and the Earth create magnetic fields
- how to describe magnetic fields by drawing lines to show their strength and direction
- that stronger magnets a smaller distance away cause a greater force on magnetic objects in the field
- that two 'like' magnetic poles repel and two 'unlike' magnetic poles attract
- how to draw field lines from the northseeking pole to the south-seeking pole
- the definitions of the terms magnetic force, permanent magnet and magnetic poles

Application

- how to explain how the direction or strength of the field around a magnet varies, using the idea of field lines
- how to explain observations about navigation using Earth's magnetic field

Extension

- how to predict the effect of changing the rating of a battery or a bulb on other components in a series or parallel circuit
- how to predict how an object made of a magnetic material will behave if placed in or rolled through a magnetic field

4 Electromagnets

In this chapter you will learn...

Knowledge

- that a current through a wire causes a magnetic field, an effect called electromagnetism
- how to describe the factors affecting the strength of an electromagnet: the current, the core and the number of coils in the solenoid
- that the magnetic field of an electromagnet decreases in strength with distance
- the definitions of the terms electromagnet, solenoid and core

Application

- how to use a diagram to explain how an electromagnet can be made
- how to explain how to practically change the strength of an electromagnet
- how to explain the choice of electromagnets or permanent magnets for a device in terms of their properties

Extension

- how to explain good and bad points about the design of a device using an electromagnet, and suggest improvements
- how to suggest how bells, circuit breakers and loudspeakers work, from diagrams

3 Magnetism

Your knowledge objectives

In this chapter you will learn:

- that magnetic materials, electromagnets and the Earth create magnetic fields
- how to describe magnetic fields by drawing lines to show their strength and direction
- that stronger magnets a smaller distance away cause a greater force on magnetic objects in the field
- that two 'like' magnetic poles repel and two 'unlike' magnetic poles attract
- how to draw field lines from the north-seeking pole to the south-seeking pole
- the definitions of the terms magnetic force, permanent magnet and magnetic poles

See page 33 for the full learning objectives.

Key word

A **permanent magnet** is a magnet all the time.

» Transition: Detecting forces

Humans are generally considered to have five senses. Each sense collects information about the outside world. But humans can't detect everything. Some things can't be seen, heard, felt, smelt or tasted – but they are still real. Scientists use measuring devices to detect things that their senses do not.

Magnetism is an example of a non-contact force. There is an attractive force between a **permanent magnet** and objects made of iron, nickel or cobalt. Steel is mostly iron so is often affected too.

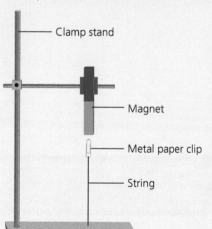

- Clamp stand
- Magnet
- Metal paper clip
- String

▲ The force holding up the paper clip is invisible

Common error

Not *all* metals are magnetic; just iron, nickel, cobalt and some of their alloys.

▲ The iron filings are moved by the magnetic force

Key words

An **attractive force** pulls things together.
A **repulsive force** pushes things apart.

Every magnet has an area around it in which magnetic materials are attracted. This is called the magnetic field and can be big or small depending on the strength of the magnet. The field can't be seen without help, but the *effect* of the field can be seen.

If two magnets are put close together, something interesting happens. Sometimes an **attractive force** pulls them together, just as we would see with a magnet and a piece of magnetic material. But sometimes there is a **repulsive force** and they are pushed apart.

It turns out that any magnet, no matter how big, has two sides, or poles. We call them the north and south poles because of what they are attracted to. Investigating the attraction and repulsion gives a simple set of rules. Magnetic materials are attracted to *both* poles of a magnet.

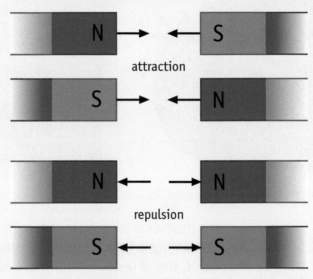

▲ Opposite poles attract each other, similar poles repel

- A north pole and a north pole **repel** each other.

- A south pole and a south pole **repel** each other.

- A north pole and a south pole **attract** each other.

Key fact

The further away an object is, the smaller the magnetic force on it.

Know >

1 Which of these objects would be attracted to a permanent magnet?

a) silver ring   **d)** human hair

b) wooden ruler   **e)** copper-nickel coin

c) iron nail   **f)** plastic bag

2 What are the three magnetic elements?

3 Give an example of a magnetic alloy.

Apply >>

4 A large iron ball bearing is put down 5 cm from a magnet and rolls towards it. The second time, it is put down 10 cm away and nothing happens. Who is closer to being right? How could you improve their answer?

5 Explain what happens in these cases when the objects are placed together (you may find the words **attract** and **repel** useful):

a) north pole and south pole

b) south pole and sample of iron

c) south pole and south pole

d) north pole and north pole

e) north pole and sample of copper

» Core: Navigation

For centuries, people have used magnetism to help them find their way. Long before GPS and mobile phones, small magnets were used to give a consistent direction. Centuries ago these were called lodestones, from an old word *lode* meaning 'journey'. They are small pieces of weakly magnetic iron ore. Modern magnetic compasses work in the same way and are used by mountaineers, sailors and other explorers to find their way.

▲ (a) Lodestones have been found by archaeologists. (b) A modern compass uses the same idea

If a magnet is allowed to move freely – perhaps by floating on water or hanging on a piece of string – it will line up with other magnetic fields. This is because of the forces of attraction and repulsion. The north pole of a magnet is the one that points towards the geographic North Pole of the Earth. It turns out that the Earth acts as if there is a giant bar magnet in the middle, with the poles at the 'top' and 'bottom' of the sphere.

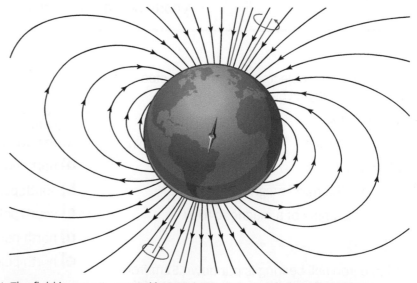

▲ The field is actually caused by moving molten iron below the surface

Common error ❗

The 'north pole' of a magnet is the end that is attracted to the (geographic) North Pole of the Earth – which is actually a magnetic south pole!

Although humans have to use equipment, some animals can detect magnetic fields with their natural senses. Many herding animals line up towards the Earth's magnetic north pole, although scientists don't know how they detect it or why they do this. Migrating animals including many species of birds, fish and even some bats use magnetic fields to find their way. Scientists have even worked out that robins have special cells in their right eyes which mean they can see the fields.

▲ Many animals have a magnetic sense

It has been suggested that humans could upgrade their senses using technology. Using a magnetic sensor, similar to those in mobile phones, in a piercing or implant would provide feedback when a person was facing north. Over time they would recognise this feeling automatically, always knowing which way was which without a compass. Other people point out the same thing can be done using augmented reality, without permanent changes to the body.

Know >

1 Why does one end of a lodestone always point north?

2 Give an example of a person who might use a compass to find their way.

Apply >>

3 Why is it important that the magnetic needle in a compass isn't heavy?

4 How would a magnetic sense be useful for a goose that migrates to Britain from the Arctic in winter?

Extend >>>

5 Give arguments for and against an implant that tells a person which way is north.

6 Should people be able to choose to have this kind of implant? Who should decide whether it is a good idea or not?

» Core: Magnetic fields

The effect of any permanent magnet on an iron nail is stronger when it is close. As the nail gets further away, it experiences a weaker attractive force. The same effect is seen between two magnets; we just need to remember that in this case it can be attractive (pulling) or repulsive (pushing), depending on the combination of poles.

If compasses – small magnets that are free to move – are placed near a permanent magnet, the direction of the field can be worked out. The patterns seen are similar to those of the iron filings. Lines can be drawn, starting at one pole of the magnet and ending up at the other. If two permanent magnets are used the lines make more complicated patterns, but they still go between opposite poles.

▲ The patterns seen with opposite poles

The closer together the lines are, the stronger the field. This shows that the field is strongest near the poles. These field lines are one way of showing the field, which is always there, but invisible. Scientists draw them with an arrow which points *from* the north-seeking pole *to* the south-seeking one.

Stopping magnetism

As distance from a magnet increases, the strength of the field decreases. Eventually it is too small to measure. There is another way to stop a magnet having an effect. A gap of air obviously makes no difference. The same is true for layers of paper or card – think about a school letter held to a fridge door with a magnet. Stronger magnets can work even through thick wood, copper or plastic.

Common error

!

Students sometimes talk about a *bigger* magnet when they mean a *stronger* magnet.

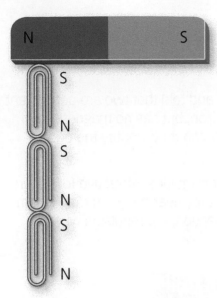

▲ The magnet induces magnetism in the paper clips

However, some materials do stop the field and so no force acts on the other side of the barrier. These materials are iron, nickel and cobalt. They are the same metals that can be permanent magnets. All of them can also be made attractive, simply by touching a permanent magnet. This is called induced magnetism and wears off as soon as they are separated. A piece of magnetic material can be given its own magnetic field if a permanent magnet is rubbed along it, steadily and slowly.

Steel can also be made into a permanent magnet. Because it is not pure iron the effect does not wear off so quickly. In fact, steel can be turned into a permanent magnet using electric currents (see Chapter 4 for details).

Know >

1 Draw the magnetic field around a bar magnet. Label where the field is strongest.

2 A field line connects two poles. Which pole is the arrow pointing away from?

3 What colour is used for the north-seeking pole of a compass or magnet?

Apply >>

4 Using paper clips and a ruler, how would you compare the strength of two magnets?

5 Explain which would make a better compass: an iron nail or a steel needle.

Extend >>>

6 Do some wider reading to find out the difference between the Earth's magnetic poles and the geographic poles.

» Extend: Magnetic particles

A student is given three metal rods and told that two are permanent magnets while the third is made of iron, but has no magnetic field. It is still attracted to the others. How can they identify the odd one out?

This example demonstrates an important idea. Attractive forces are the first thing people think of when they hear the word magnetism. The true test of a magnet is not attraction, but repulsion when certain combinations are used.

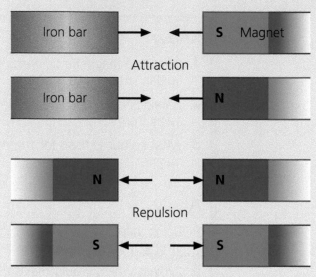

▲ The iron bar is attracted no matter which way around, but repulsion is only seen with the magnets

Key word

An **electromagnet** is a non-permanent magnet turned on and off by controlling the current through it.

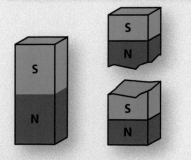

▲ Each piece has its own north- and south-seeking poles

Repulsion between very powerful magnets – often an **electromagnet**, as explained in Chapter 4 – is used to reduce contact between moving surfaces and so minimise friction. This means vehicles can move very quickly, for example magnetic levitation or *maglev* trains.

Causes of magnetism

No matter how many times examples are seen, there is something about demonstrations of magnetism that look magical. Even though the non-contact force of gravity is also invisible, holding something *up* with magnetism surprises people more than objects falling.

To understand why some materials are magnetic, scientists start by breaking magnets into smaller pieces. It turns out that cutting a magnet in half makes two smaller, slightly weaker magnets. Cutting each of those pieces in half gives even smaller magnets. In fact, there is no piece of iron too small to be a magnet.

Key word

An **atom** is the smallest possible part of an element.

A good way to think about it is that each **atom** of iron behaves like a tiny, very weak magnet. Even a small piece of iron will contain so many iron atoms that the effect is bigger. Most of the time, these magnets all point in different directions and cancel out. But if they line up, the overall forces of attraction and repulsion add up and can be measured. This is why any magnetic material can be turned into a magnet, because the tiny magnets can be made to point in the same direction.

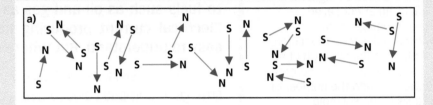

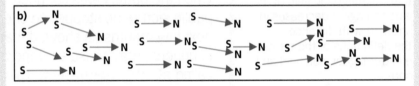

▲ (a) The particles in a piece of unmagnetised iron. (b) Particles in a permanent magnet

Tasks

1. Why is repulsion the true test of magnetism?
2. A child's toy train has three parts which stick together with magnetism. To save money, the manufacturer makes the middle part with iron rather than permanent magnets. Draw a diagram to show what happens.
3. Why don't we use maglev trains everywhere? What disadvantages can you find out about?
4. The examples above talk about iron, ignoring cobalt and nickel. Research why we use iron in most cases where a magnet is needed.

4 Electromagnets

›› Transition: Moving magnets

Your knowledge objectives

In this chapter you will learn:
- that a current through a wire causes a magnetic field, an effect called electromagnetism
- how to describe the factors affecting the strength of an electromagnet: the current, the core and the number of coils in the solenoid
- that the magnetic field of an electromagnet decreases in strength with distance
- the definitions of the terms electromagnet, solenoid and core

See page 33 for the full learning objectives.

Key word

Electromagnetism is the temporary magnetic field caused by an electrical current in any material.

One of the important ways electrical devices are used is to produce movement. Batteries provide the only real alternative to fuels such as oil and gas as a way to make things move. Electrical current producing heating and lighting effects is easy to understand. But where does a force come from?

In 1820 a physicist called Hans Oersted noticed something strange. A compass placed near an electrical circuit wobbled when the battery was connected. Despite being in the middle of a public lecture he stopped to check what was happening. Over the next few months he showed that when charges move through any conductor, they cause a temporary magnetic field around it. This is called **electromagnetism**. There are three possible effects.

- Other magnets (whether permanent or other electromagnets) will be attracted or repelled, depending on direction.

- Anything made of magnetic materials – iron, nickel or cobalt – will be attracted.

- Other objects and materials do nothing.

These follow the same rules as permanent magnets, as explained in Chapter 3.

Common error

Unlike permanent magnets, an electromagnet does *not* have to be made of the three magnetic materials.

▲ When current flows, the compass needles move, showing the electromagnetic field

An electromagnet is often used in situations when we want to be able to turn the magnetism on and off. Common examples include sorting recycling, securing vault doors and medical imaging.

Clearly all magnetic fields cause forces. Whether these forces lead to movement depends on the situation. This effect is used in many devices, large and small, to control movement with an electrical circuit.

▲ All of these devices use an electric current to make an electromagnetic field, which leads to movement

Know >

1 How can an electromagnet be turned off?

2 Which combinations of metals could be separated with an electromagnet?

a) iron and steel

b) steel and aluminium

c) iron and copper

d) copper and aluminium

3 Give an example of a household device that uses electromagnetism to cause movement.

Apply >>

4 How can a copper wire cause a magnetic field?

5 Why do walkers avoid using their compasses near electricity pylons?

Key fact

Like a permanent magnet, the strength of an electromagnet decreases with distance.

» Core: Stronger electromagnets

A single wire with a small **current** doesn't make a very strong magnetic field. It's not hard to predict some ways to make it stronger and so more useful.

Increasing the current makes the field stronger. This is done by increasing the **potential difference** from the supply – for example by using more cells – or by reducing the **resistance** in the circuit. High currents can be dangerous because the wire can become hot.

A simpler way is to wrap the wire in a spiral or **solenoid** so that the small effect from each part of the wire adds up. This means the current in the wire can be kept low but the electromagnetic field is stronger. The wires must be covered in an insulating material.

Because of the shape, a solenoid has a similar magnetic field to a permanent bar magnet. One end of the coil is a north-seeking pole, the other is a south-seeking pole. This electromagnetic field is only there while the current is flowing. If the current flows in the opposite direction (e.g. by turning the cell around) then the poles are reversed.

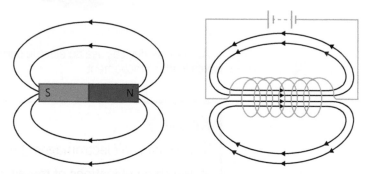

▲ The two ends of the coil behave like the two ends of a bar magnet

The final way of making a stronger electromagnet is harder to predict. It turns out that if we put a piece of magnetic material – iron, nickel or cobalt – in the centre of the solenoid, the magnetic field is stronger. Pure or 'soft' iron is normally used to make this **core** because it works best. It is also cheaper because it is more common.

To make the strongest electromagnets, a high current is passed through cables wrapped many times around a soft iron core. The magnetic field can easily be millions of times stronger than that of the Earth.

Key words

The flow or movement of charge is called **current** and is measured in amperes (A).

$$\text{resistance } (\Omega) = \frac{\text{potential difference (V)}}{\text{current (A)}}$$

$$R = \frac{V}{I}$$

Resistance measures how hard it is to push charges through a material or component. It is measured in ohms (Ω).

A **solenoid** is a wire wound into a tight coil, part of an electromagnet.

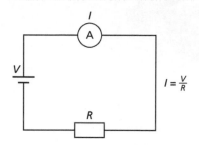

$$I = \frac{V}{R}$$

▲ The current (*I*) that flows is equal to the potential difference (*V*) divided by resistance (*R*)

Key word

The **core** of an electromagnet is a piece of soft iron which the solenoid is wrapped around.

Common error !

The iron core is not part of the electrical circuit because it is not touching the wire.

▲ To get a stronger field we could increase the current or number of coils of wire in the solenoid. We could also put a piece of soft iron in the centre

Know >

1 Give two ways to increase the strength of an electromagnet.

Apply >>

2 The top end of an electromagnet acts like a north-seeking pole when current flows. How could it be made to act like a south-seeking pole?

3 A steel ball bearing is placed close to a solenoid. The current in the wire is turned on, then off. Describe what happens to the position of the ball bearing with each action.

Extend >>>

4 Which is more dangerous – high potential difference or high current? Explain your answer.

5 An engineer suggests increasing the strength of an electromagnet by using a longer wire so it can have twice as many coils. The problem is that it will then have double the resistance too. Can you explain why it won't be any stronger?

Key facts

Increasing the potential difference *increases* the current. Increasing the resistance in the circuit *reduces* the current.

High potential differences cause shocks and sparks; high currents cause heating and can kill.

▲ An electromagnet attracts iron and steel materials but leaves aluminium behind

Key word

The **interval** is the gap between values of a variable.

» Core: Using electromagnets

Sometimes electromagnets are used simply because they can be turned on or off. It is also easy to change the strength of the electromagnetic field. A common example is sorting aluminium from steel waste – it is no good picking up the steel parts if they can't be released. At some point the steel must 'unstick' so it can be recycled!

Other common examples rely on the fact that current can be increased or decreased quickly and accurately. This means the strength of the field – and the force acting – can be changed with a very small **interval**. This will happen as quickly as the current changes, which can be controlled in fractions of a second.

Making sound

A basic loudspeaker uses two opposing magnets, one of which is permanent and the other a controllable electromagnet. A solenoid is usually stuck to the back of the loudspeaker cone. The case includes a permanent magnet, lined up with the coil of the electromagnet. The amount of attraction or repulsion depends on the direction and strength of the electromagnetic field – which depends on the direction and size of the current 'driving' the loudspeaker.

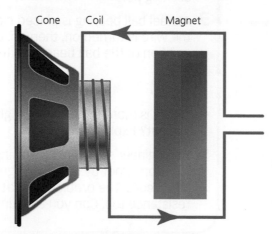

Cone   Coil   Magnet

▲ The movement of the loudspeaker causes a sound in the air

1 The current in the solenoid increases.

2 This increases the strength of the electromagnetic field.

3 This means there is more attraction between coil and permanent magnet.

4 The cone moves back *towards* the permanent magnet.

Key words

The number of complete waves detected in one second is called the **frequency**.

Amplitude Amplitude is the height of a wave, measured from the middle.

★ **See Pupil's Book 1, Chapter 7, for more information.**

If the direction of the current is reversed, there is more repulsion instead of more attraction and the cone moves forward, *away* from the permanent magnet.

The rapid movement of the cone back and forth causes the air to vibrate. This is what we recognise as sound. The **frequency** and **amplitude** of the movement determines the pitch and loudness of the sound.

▲ The faster the cone vibrates, the faster the air vibrates, so the higher the pitch

All loudspeakers use this process to convert an electrical signal to sound. Stronger but smaller permanent magnets, often made using rare earth metals such as neodymium, mean speakers can be small but loud.

Know >

1 What would the crane operator in the photo on the opposite page, do if the electromagnet couldn't lift a piece of steel?

Apply »

2 An earlier example of electromagnet applications was a vault door. Why might bank robbers cut off electrical power to the building?

3 Why do you think most loudspeakers have the solenoid rather than the permanent magnet on the moving cone?

Extend »»

4 Briefly research magnets made of materials other than iron, nickel and cobalt. Summarise their properties.

» Extend: Sound and motion

A loudspeaker needs to be able to make a wide range of sounds – high-pitched and deep, loud and quiet. This is done by small, rapid changes to the current supplied to the electromagnet. An electric bell also uses electromagnetism, but in a simpler way; there are examples in or near most classrooms. The design of the electric bell is nearly 200 years old but because of the reliability it is still popular for uses such as fire alarms.

Instead of changing strength and direction, the current is simply turned on and off. If a bulb is in the circuit it might be seen to flicker. This means the electromagnet is also turned on and off quickly, opposing the spring.

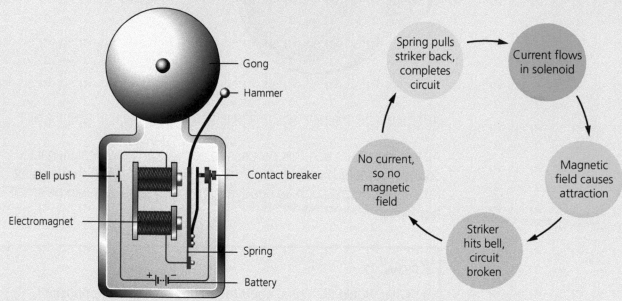

▲ The electromagnet is turned on and off quickly

▲ The vibrating striker causes the bell to ring

All of the examples so far have involved movement back and forth. This is ideal for causing vibration and sound. A different approach is used to cause steady movement, for example a spinning fan or the wheels driving an electric car. The electric motors in these applications still use attraction and repulsion between permanent and temporary magnets.

Relays

Large currents are dangerous because they cause heating and damage. Large potential differences are dangerous because they can make a spark across a gap. This means that switching a large current on and off can be difficult.

A relay is the name of a device that solves this problem. Like an electric bell, a small current flows in a circuit when the switch is closed. But instead of moving a piece of metal to hit a bell, a separate switch is moved, which completes a second circuit.

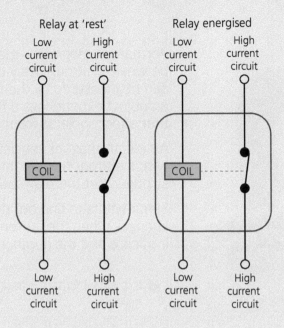

▲ The low and high current circuits are not connected electrically

Relays like this are used in vehicles, so that the ignition key completes a safe circuit rather than being connected to the car battery.

Common error

When explaining relay switches, make clear which circuit you are referring to.

Tasks

1. How would you change the electrical signal to a bell if you wanted the sound to be:
   a) louder?
   b) higher-pitched?
2. The 'key' circuit in the relay switch of a car has a small cell. What difference will it make if this only supplies half the intended current?
3. Create a flow chart like the one opposite for an electrical relay.
4. Research a simple electric motor. What similarities and differences are there in different devices?

Enquiry:
Making stronger magnets

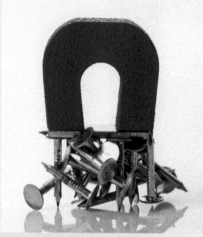

▲ The strength of a permanent magnet can be measured but not changed

Permanent magnets – made of iron, nickel, cobalt or steel – can be weak or strong. They can't be turned on or off. The strength can be measured by the mass they can lift, or by how far away an object experiences a force. Stronger magnets have a field that affects objects from further away.

An electromagnet consists of a coil of wire, called a solenoid, with a current flowing through it. A soft iron core makes it stronger, but there are other ways to change the strength too.

Samir wants to find out how the strength of an electromagnet changes when the current is increased. Ella asks him why the iron core and the number of coils will need to be kept the same.

① Explain why this is important. What are these kinds of variables called?

The current in the electromagnet will be the independent variable. Samir thinks he should use 'nails picked up' as the dependent variable but Ella tells him that paper clips would be better. She says the reading would be more sensitive.

② What does she mean?
③ A variable resistor is used to change the current in the electromagnet from 0.1 A to 0.9 A. What is the range?
④ Ella says 'My theory is that more current will mean more paper clips can be picked up.' How could this sentence be improved?

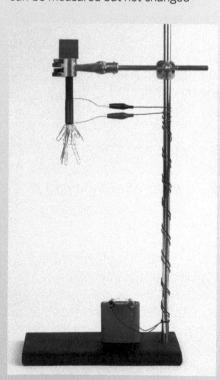

▲ Samir and Ella test the equipment

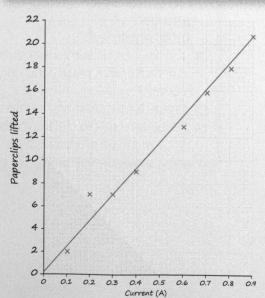

◄ Samir has drawn a best fit line from the data

5 Describe the pattern shown by the graph.

One point was missed during the experiment, and one was ignored when the line was drawn.

6 No data was recorded for a current of 0.5 A. How many paper clips would you expect the electromagnet to pick up for this current?

7 Explain why the reading at a current of 0.2 A was ignored.

In the next lesson, Ella repeats the experiment investigating how the strength of the electromagnet depends on the current in the wire. The solenoid she uses has a different number of coils but she keeps it the same through the experiment.

8 Describe the pattern you would expect her to see and why.

In a recycling plant, cans are sorted by using an electromagnet. This separates steel and aluminium cans, which need to be recycled in different ways. The chief engineer finds that some steel cans are not being lifted by the electromagnet as expected.

▲ Steel and aluminium cans ready to be separated by an electromagnet

9 Suggest two changes they could make that would make sure the cans are separated properly. Which is better from a business point of view?

Energy

Learning objectives

5 Work

In this chapter you will learn...

Knowledge

- that work is done and energy transferred when a force moves an object
- how to describe how force and distance affect work done

Application

- how to draw a diagram to explain how a lever makes a job easier
- how to compare the work needed to move objects different distances

Extension

- how to use the formula: work done (J) = force (N) × distance moved (m) to compare energy transferred for objects moving horizontally
- how to compare and contrast the advantages of different levers in terms of the forces need and distance moved

6 Heating and cooling

In this chapter you will learn...

Knowledge

- how to give the direction of energy transfer between objects of different temperature
- the definitions of the terms thermal conductor and thermal insulator
- how to name, and describe, the three pathways by which thermal energy is transferred

Application

- how to explain observations about changing temperature in terms of energy transfer
- how to describe how an object's temperature changes over time when heated or cooled
- how to explain how a method of thermal insulation works in terms of conduction, convection and radiation
- how to sketch diagrams to show convection currents in unfamiliar situations

Extension

- how to sketch a graph to show the pattern of temperature change against time
- how to evaluate a claim about insulation for clothing technology
- how to compare and contrast the three ways that energy can be moved from one place to another by heating

5 Work

» Transition: Forces and energy

Forces can change the motion of objects. Although we can't see the forces themselves, we can see the effect they have.

If we look at someone pushing a sledge, then we can draw a force diagram which shows the forces acting on the sledge.

Your knowledge objectives

In this chapter you will learn:
- that work is done and energy transferred when a force moves an object
- how to describe how force and distance affect work done

See page 53 for the full learning objectives.

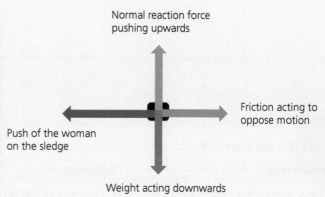

▲ (a) Having fun sledging in the snow. (b) Forces acting on the sledge

Key word

Work is the transfer of energy when a force moves an object in the direction of the force. Work is measured in joules.

The push force from the woman is greater than the force of friction, so there is a **resultant** force on the sledge. It will speed up.

The woman's store of chemical energy will decrease. The sledge's store of kinetic energy will increase. Some energy will also be dissipated due to friction, increasing the thermal energy store of the surroundings.

There is transfer of energy due to the action of the push force.

This transfer of energy is called **work**. Work is measured in joules.

Common error

In everyday life, we use work to talk about different situations. In science, work has a very specific meaning. Work is done when an applied force causes an object to move in the direction of the force.

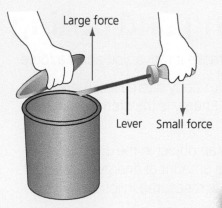

Large force

Lever   Small force

▲ Using a screwdriver as a lever to open a tin of paint

Levers

We can use a lever to make doing work easier. A lever is a type of simple machine. In the diagram, the screwdriver is being used as a lever.

To open a tin of paint, a screwdriver is put under the tin lid. You push down with a small force on the handle. The screwdriver pivots on the rim of the tin, and a larger force acts upwards. The amount of work done by each force is the same.

Worked example

Annie is working out in the gym. She is doing bicep curls. She lifts the dumbbell up about 20 cm. Explain why work is being done by Annie.

To lift the dumbbell, Annie has to apply a force. The force is in an upwards direction (against gravity). The dumbbell moves upwards in the direction of the force. Therefore, work has been done by Annie to lift the dumbbell.

Apply ≫

1 Give two examples of situations in which you do work.

2 The store of gravitational potential energy of the dumbbell has increased. Identify which energy store has emptied.

3 Give another example in which a lever can be used to make work easier.

» Core: Pushing and pulling

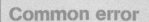

When you push or pull a box and make it move, you apply a force.

The store of chemical energy in your body decreases. The kinetic energy store of the box increases. Energy is transferred and you have done work.

Work is done when a force moves an object in the direction of the force. Work is a measure of the amount of energy transferred between stores. Like energy, work is measured in joules.

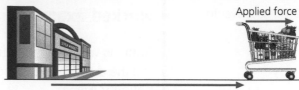

Distance moved

In the picture above, the trolley moved a distance in the direction of the applied force. Work was done.

The bigger the force, the more work is done. If you double the force, the work is doubled.

The further the object moves in the direction of the force, the more work is done. If you double the distance, the work is doubled.

Work is also done when an object is deformed. This **deformation** can be due to compression or tension forces.

Astronauts on board the International Space Station (ISS) have to do a lot of exercise to keep themselves healthy. Their bodies do not experience gravity in space, and so bones tend to become weaker and more brittle.

The astronauts run on a treadmill. Bungee cords are used to keep the astronaut in contact with the treadmill. Standing up straight stretches the bungee cords. The bungee cords pull downwards on the astronaut, simulating the effect of gravity. This helps to strengthen bones and muscles. The astronaut is doing work on the bungee cords because he is deforming the bungee cords.

Common error

If you push hard against a wall, you are applying a force. However, if the wall doesn't move, then you are not doing work. Work is only done when the object moves in the direction of the force.

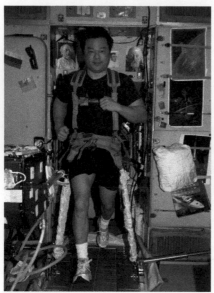

▲ Astronaut Leroy Chiao, exercising on the International Space Station. A bungee cord harness helps make the exercise more effective

Worked example

Xander pushes his trolley. His brother puts some toys in the trolley. Xander now has to push twice as hard to move the trolley the same distance.

Describe how the work done moving the trolley changes.

The force has increased. The distance has stayed the same. The work done will double because the force has doubled.

Apply »

1 Mari does 150 J of work lifting a box up 0.5 m. How much work would Mari do if she lifted the same box up 1 m?

2 A builder puts 14 breezeblocks in a wheelbarrow. He pushes the wheelbarrow a distance of 100 m in a straight line. He then pushes the wheelbarrow a further 500 m. Describe how the work done by the builder changes between the two sections of his journey.

Extend »»

3 For each of the following situations, explain if work is being done.

a) A crane is used to lift a load of gravel.

b) A fridge magnet is attached to the fridge by a magnetic force.

c) You climb up a set of stairs.

d) You read a book.

▲ Using rollers to make it easier to move a very heavy object

▲ Early wheels were probably made from wood

» Core: Simple machines

Levers, pulleys and wheels are all examples of simple machines. A simple machine is a device that is used to make work easier.

Wheels

The earliest 'wheels' were probably logs, which were used as rollers. This reduces the friction which opposes motion. The resultant force is larger and this makes it easier to move an object.

The problem with rollers is that you have to keep moving them as the object moves. Putting the wheel on an axle meant that the wheel could rotate as the object moved.

Pulleys

A pulley is a wheel that has a groove around the edge. A rope sits in the groove. Using a simple pulley makes it more convenient to lift a heavy weight because people find it easier to pull down on a rope that lifts up the heavy weight. Often, more than one pulley is used at a time, which makes it easier to do work.

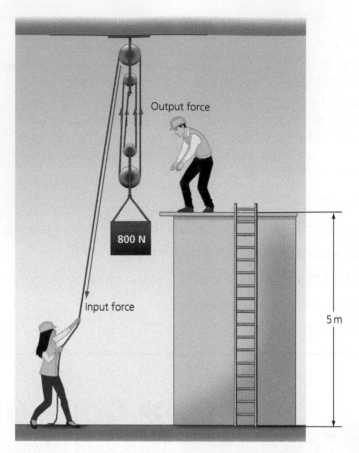

▲ A simple pulley

▲ Using a pulley reduces the force needed to lift a weight

Key words

The **input force** is the force you apply to the machine.

The **output force** is the force that is applied to the object being moved by the machine.

Displacement is the distance an object moves from its starting point as measured in a straight line from that point.

In the diagram above there are four pulleys. The **input force** is four times smaller than the **output force**. However, each person pulling on the rope has to pull four times as far as the **displacement** of the object that is being lifted.

Levers

A lever is a rigid bar which pivots about a point, known as a fulcrum.

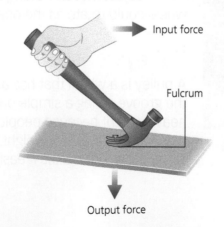

When a claw hammer is used to pull out a nail, the nail is the fulcrum. The input force moves a greater distance than the output force.

Levers and work

We can calculate the work done by the input and the output force. This is equal to the energy transferred by each force.

Energy is conserved in the system. That means that the work done by the input force must be the same as the work done by the output force.

This means that the input force is smaller than the output force because the distance the input force moves is larger than the distance the output force moves.

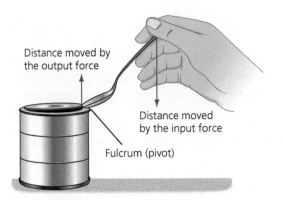

Distance moved by the output force

Distance moved by the input force

Fulcrum (pivot)

Worked example

Tess is moving a boulder with a lever. Tess applies an input force of 100 N on the end of the lever. When she does this the end of the lever moved down a distance of 50 cm. The other end of the lever moves up a distance of 10 cm.

What is the largest weight of boulder that Tess can move with the lever?

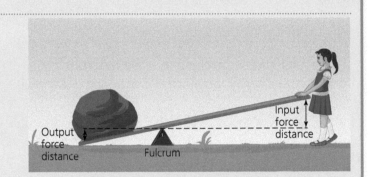

The input force distance is 5 greater than the output force distance. The work done by each force will be the same. So, the input force will be 5 smaller than the output force.

maximum output force = 5 × input force = 5 × 100 N = 500 N

Tess will be able to move a boulder with a weight of 500 N.

Know >

1 What is the unit for work?

2 Explain what is meant by 'resultant force'.

Apply >>

3 Tess now applies an input force of 200 N on the end of the lever. What is the largest weight of boulder that Tess can move with the lever?

4 Explain how a small child could lift an adult off the ground if they had a long, strong plank of wood.

Extend >>>

Your forearm is an example of a lever. This type of lever is a **distance-multiplying lever**. The input force is due to your bicep muscles contracting. The output force is the force required to lift the object in your hand against the force of gravity.

5 Sketch a diagram to show how your forearm is a distance-multiplying lever.

Enquiry >>>>

Levers can be classified into different types depending on where the fulcrum is. In a first-order lever the fulcrum is between the input and the output forces.

6 Research the different types of levers and give examples of their use.

» Extend: Calculating work

Strongest women and strongest men competitions often include a 'truck pull' event. In this event the competitors wear a harness that is attached to a large vehicle. They then pull the vehicle a set distance.

The amount of work done to move the vehicle can be calculated using the equation:

work done (J) = force (N) × distance moved in the direction of the force (m)

Worked example

In the picture Kathleen is pulling a truck with a force of 72,000 N.
The truck moves a distance of 20 m.

a) Calculate the work done to move the truck.
b) State the energy transferred during the truck pull.

▲ Kathleen is doing work as she pulls the truck

a) work done = force × distance moved
   = 72,000 N × 20 m = 144,000 J

b) Work done is the energy transferred, so energy
   = 144,000 J

Tasks

❶ Competitors in strongwomen competitions start training by pulling smaller vehicles. Complete the table to show the work done by one competitor pulling different vehicles in training.

Vehicle	Force needed to move vehicle (N)	Distance moved (m)	Work done
Small car	7 400	10	
Small car	7 400	20	
Medium car	11 640	10	
Large car	24 000	10	

❷ Another strength training exercise is the sled pull. The weight that a person can move using a sled is smaller than the weight they can move in a vehicle pull. Use your knowledge of simple machines, forces and work done to explain why it's harder to move the sled.

▲ An athlete training for the sled pull

Uses of levers

There are examples of levers in many different places. These levers can be classified depending on where the fulcrum is.

▲ An oar has the fulcrum between the input and output forces

▲ Using tweezers to handle small objects

▲ Nutcrackers can be used to apply a large force to a nut

Tasks

❸ For each of the three levers shown in the pictures, sketch a diagram to show:
   a) the fulcrum
   b) the position of the input and output forces
   c) the relative distances the input and output forces move.
❹ Use the information from your diagram to discuss the advantages of each type of lever shown.

Enquiry:
Rising high

▲ The Burj Khalifa in Dubai

Buildings with 10 or more floors have been known since around the 11th century. The weight of the structure limited buildings much taller than this.

▲ The skylines of many large cities around the world are dominated by skyscrapers

In the 1880s, the invention of steel frames to provide structure to buildings allowed taller structures to be built. Many architects designing new buildings try to build the tallest structure. The tallest building in 2016 is the Burj Khalifa in Dubai. It is 830 m tall and has 163 floors. This will be beaten by the Jeddah tower, due to open in 2020. It will be 1 km tall.

Another key invention for skyscrapers was the lift. Without a lift, no-one would want to live or work on the top floors of tall buildings.

The safety lift was developed by Elisha Otis in 1852, and the first electric lift was built in 1880 by Werner von Siemens.

Many lifts have a large motor at the top of the building. This is used to pull the lift up. Work is done by the motor to pull the lift up – the diagram shows the forces acting on the lift.

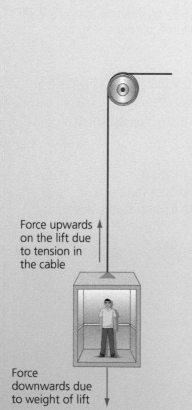

Force upwards on the lift due to tension in the cable

Force downwards due to weight of lift

❶ Describe the relative sizes of the forces when:
   a) the lift is starting to move
   b) the lift is moving at a steady speed.

Energy transfers in the lift system

In a skyscraper, the lift motor turns a pulley which pulls the cable upwards. There is a force of tension. This lift moves upwards through different heights.

The lift's store of gravitational potential energy increases as it rises.

The energy is transferred by the electric motor. The electric motor does work:

work done (J) = force (N) × distance moved in the direction of the force (m)

Two students want to model the lift system to compare the energy transferred by the motor and the energy gained by the lift. They have a small motor, a pulley, some masses and a power supply. They set up their equipment as shown below.

▲ Pulley and cables control the movement of the lift

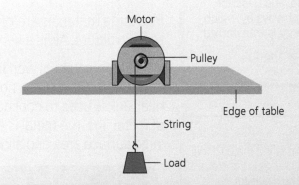

The students use the motor to lift a load of 250 g through a height of 1.5 m.

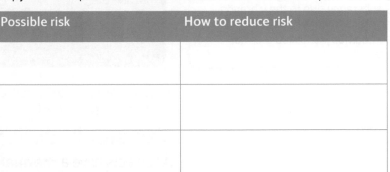

❷ Copy and complete the risk assessment table for the experiment:

Possible risk	How to reduce risk

❸ Calculate the weight of the load.
❹ Calculate the work done lifting the load.
❺ Explain why the motor has done work. When the students calculate the electrical work done by the motor, they find that it is 60 J.
❻ Explain what has happened to the additional energy transferred by the motor.

Heating and cooling

Your knowledge objectives

In this chapter you will learn:
- how to give the direction of energy transfer between objects of different temperature
- the definitions of the terms thermal conductor and thermal insulator
- how to name, and describe, the three pathways by which thermal energy is transferred

See page 53 for the full learning objectives.

Key words

A **thermal insulator** is a material that reduces how fast energy is transferred between hot and cold objects.

A **thermal conductor** is a material that increases how fast energy is transferred between objects.

A **thermal store of energy** is a measure of the energy stored in a substance due to the vibration and movement of particles. It is sometimes just called thermal energy.

» Transition: Thermal insulators and conductors

People who work in cold environments often wear jackets and trousers made from a material that is a thermal insulator. Homeowners are encouraged to insulate their lofts. This helps to reduce energy transfers from the warmer inside of the house to the colder outdoors.

Insulating materials are often made so that they trap a layer of air in their structure. Air is a very good thermal insulator.

When a computer is working, the central processing unit (CPU) gets very warm. If it gets too hot the computer will break. A device called a heat sink is used to cool down the CPU. A heat sink is simply a series of metal fins. The metal is a **thermal conductor**. The fins have a large surface area and allow the CPU to cool down more quickly.

▲ Loft insulation is used to reduce the waste of energy in houses

▲ A computer motherboard showing the aluminium heat sink

Transfer between thermal energy stores

All objects have a **thermal store of energy**. Hotter objects have a larger thermal store of energy than colder ones.

When there is a temperature difference between objects that are in contact with each other, the warmer object will transfer energy to the cooler object.

The warmer object gets cooler, and the cooler object gets hotter. Energy is transferred between them.

An insulator will slow down the energy transfer. The warmer object will stay at a higher temperature for longer.

Worked example

A student is measuring how well polystyrene cups keep a drink warm. She uses three different combinations of cups stacked inside each other.

The student puts the same amount of hot water in each cup. She measures the temperature of the water in each cup every 10 minutes. Her results are presented in the table.

| A: 1 cup | B: 2 cups | C: 3 cups |

1 Calculate the temperature drop after 20 minutes for each cup.

2 Which combination of cups was the best thermal insulator? Use data from the results table to explain your answer.

	Temperature (°C) after		
	0 minutes	10 minutes	20 minutes
A: 1 cup	70	48	39
B: 2 cups	70	54	44
C: 3 cups	70	60	55

1 Temperature = initial − final
  change   temperature   temperature

A: Temperature change = 70 °C − 39 °C
                      = 31 °C

B: Temperature change = 70 °C − 44 °C
                      = 26 °C

C: Temperature change = 70 °C − 55 °C
                      = 15 °C

2 Combination C (three cups together) was the best thermal insulator. After 20 minutes the temperature had only dropped by 15 °C, which was the smallest temperature drop.

Know >

1 Decide whether these materials are thermal insulators or thermal conductors.

- Aluminium
- Bubble wrap
- Copper
- Polystyrene
- Wood

Apply >>

Some students are investigating the thermal properties of materials. They have: newspaper, foil, cling film, bubble wrap, four identical plastic cups and thermometers. They wrap the sides of each cup in a different material. They put the same amount of liquid into each cup and measure the temperature every 10 minutes. Their results are shown in the table.

	Temperature of liquid after …. (°C)		
	0 minutes	10 minutes	20 minutes
Bubble wrap	70	64	51
Cling film	70	56	29
Foil	70	57	46
Newspaper	70	52	26

2 Explain why the students used identical cups and put the same amount of liquid in each cup.

3 What safety precautions should the students have used during their investigation?

4 Which material is the best thermal insulator? Use data from the table to explain your answer.

Common error

You might have heard people say 'close the door, you'll let the heat out'. Heat isn't a physical substance, so it can't move anywhere. What they mean is that you'll set up a temperature difference between the two places. This makes the warmer air in the room start to move outside, and the colder air outside start to move inside.

» Core: Melting moments

Put a hot drink in contact with a cold ice cube. The hot drink cools down, but the ice hasn't given 'coldness' to the drink. Instead, there is a transfer of energy from the thermal store of the drink to the thermal store of the ice.

The temperature difference between the ice cube and the drink leads to a transfer of energy between them until the **temperature** of each is the same.

Particle model

All substances have a thermal store of energy. This is because the particles are always in motion.

In a **fluid**, particles can move around easily. In a solid, the particles can't change position easily, but they do vibrate more.

Particles in a warmer fluid will be moving faster than the particles in a colder fluid. Particles in a warmer solid will vibrate more than the particles in a colder solid.

> **Key word**
>
> The **temperature** of an object is a measure of the motion and energy of the particles that form the object.

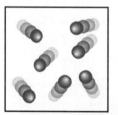 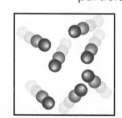

▲ In the particle model of matter, the higher the temperature of a liquid or a gas the faster the particles move

▲ When a solid is heated, the particles vibrate more

When we put a cold substance in contact with a warm substance there is a temperature difference.

The particles in each substance can now interact. When the faster-moving particles bump into slower-moving particles there is a transfer of energy. The faster particles slow down and lose energy. The slower particles speed up and gain energy.

The total energy of the system will stay the same because energy cannot be created or destroyed.

▲ Warming hands up by holding a hot drink, or cooling a hot drink down?

Worked example

Describe the energy transfers that take place as a snowman melts on a sunny day.

There is a temperature difference between the snowman and the air. The air will be warmer than the snowman. The air particles will lose energy and the snow particles will gain energy. However, the total energy of the system will stay the same.

Know >

1 Compare and contrast the arrangement and motion of particles in a solid, liquid and gas.

2 Describe what happens to the temperature of an object as its thermal energy decreases.

Apply »

3 Describe the energy transfers that take place between hot toast and butter.

4 A chef places a bowl of milk at 50 °C on a table in a room where the temperature is 18 °C. She leaves the bowl on the table for 10 minutes. The chef then puts the bowl in an oven which has been heated to 200 °C.

a) Describe what happens to the temperature of the milk in each case.

b) Explain your answer using the idea of energy transfers.

Extend »>

5 As ice melts, its temperature remains at 0 °C. However, its store of thermal energy increases while it melts. Suggest why.

Enquiry »»

A student measures the temperature of a cup of coffee for 10 minutes using a thermometer. This graph shows her results.

6 Write down the independent and dependent variables in this experiment.

7 The temperature is a continuous variable. Explain what this term means.

8 Use the graph to describe the relationship between temperature of the coffee and time.

9 Draw a sketch graph showing the thermal energy of the coffee during this time. What unit should be used for thermal energy?

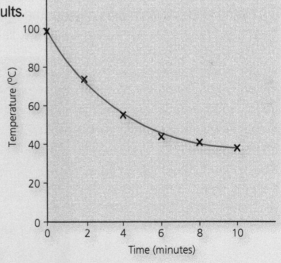

▲ Graph showing the temperature of a cup of coffee over time

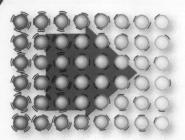

▲ Particle collisions lead to a transfer of energy by conduction

Key word

Conduction is the transfer of energy by the vibration and collision of particles.

Key word

Thermochromic film changes colour depending on the temperature. It can be used as a simple thermometer.

» Core: Conduction

Temperature differences lead to energy transfers between objects. There are three ways that this energy transfer may happen: conduction, convection and radiation.

Conduction is the main method of energy transfer in solids. As an object heats up, the particles vibrate more in their position. When the particles bump into each other, they transfer energy.

Conduction takes place easily in solids. This is because the particles are close to each other with strong bonds between them. Solids are good thermal conductors.

Some solids are better insulators than others. Wood and plastic are thermal insulators, and this is why they are used to make pan handles and cooking spoons.

Investigating thermal conductivity

You can use **thermochromic film** to investigate the difference in thermal conductivities of different materials. This is shown in the photo above. A metal coin and plastic button are put on the thermochromic film. Putting a finger on each of them for a short time, energy is transferred from the warm object to the cooler object. When the coin and the button are moved, the thermochromic paper has warmed up underneath the coin, but not the button. The button has reduced the energy transfer to the paper. It is a thermal insulator.

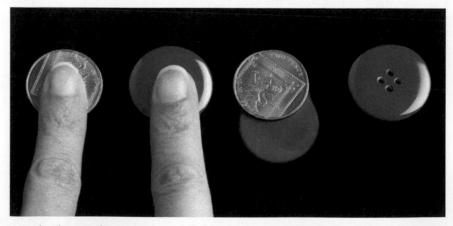

▲ A simple experiment to show different thermal conductivity in metals and plastic

Conduction in fluids

The particles in a liquid are close together, but they can move easily. Conduction can take place in liquids, but energy transfer is usually by convection.

Gases, such as air, have widely spread particles, with almost no bonds between them. This makes them very good thermal insulators.

Conduction in metals

Metals are very good thermal conductors. Metals contain free electrons that are not bound to atoms and are free to move through the metallic structure. As the metal heats up, these electrons move faster and collide more often with the particles of the metal. This makes the energy transfer quicker and more effective than solids without free electrons.

◄ Apparatus to show the different thermal conductivities of metals. A thermochromic film changes colour as the bar heats up. The better the thermal conduction, the faster the colour change

Worked example

This is a simplified diagram of a solid bar being heated.

a) In which direction is energy transferred?
b) Describe how energy is transferred along the bar.

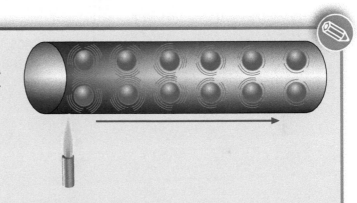

a) Energy is transferred from the end that is being heated because it has a higher temperature.
b) The particles near the flame vibrate more. They bump into the surrounding particles and transfer energy. These particles then start to vibrate more and are more likely to bump into the particles next to them. Energy is transferred between the particles and the temperature along the bar rises.

Know >

1 Write a definition for temperature.

2 What unit is energy transfer by conduction measured in?

3 What do you call a material that does not allow thermal energy to be transferred easily?

Apply »

4 A cooking pan has a copper base and a plastic handle. Explain why.

5 Houses often have insulation between the walls. How does the insulation reduce energy transfer between the inside and outside of the house?

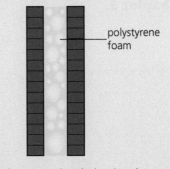

polystyrene foam

▲ Polystyrene insulation in a house wall

Extend »»

6 Quick Thaw plates are used to defrost frozen items very quickly.
They are made from a metal plate, with grooves in the surface.
Suggest how a Quick Thaw plate is able to defrost frozen items
very quickly.

Enquiry »»»

7 Standing on a tiled floor with bare feet feels much colder than
standing on a rug in the same room. Suggest why this happens,
and write a hypothesis that could be used to test this observation.

»» Core: Convection

Key word

Convection is the transfer of
energy when the particles in a
heated fluid rise.

Convection is the main method of thermal energy transfer
in fluids. In fluids, the particles are free to move, and energy is
transferred as they move.

► The hot wax is less dense than the
surrounding liquid so it floats to the top
of the lava lamp

▲ Hot air balloons float because the warm air inside the
balloon is less dense than the cold air around them

★ **You can read more about
density in Pupil's Book 1,
Chapter 9.**

When part of a fluid is heated, it expands. Its volume increases, but
its mass stays constant. This means that the volume of warmer fluid
is less dense than the surrounding cooler fluid. It will float up to the
top of the cooler fluid.

Hot air balloons make use of this property of fluids. The air inside
the balloon is heated and so it expands. The same mass of air takes
up a greater volume. The air in the balloon is less dense than the
air surrounding it. Just as a cork floats on water because it is less
dense than the water, the hot air balloon floats in the air around it.

Convection currents

The movement of particles leads to a transfer of energy between the
faster-moving particles in the warm air and the slower-moving particles
in the cold air. As the particles transfer energy they will move more
slowly. The cooling air will increase in density and sink down again.

▲ A radiator is used to heat air. This warm air circulates around the room due to convection currents

This sets up convection currents where warm air rises and sinks down as it cools.

Radiators should really be called convectors because they heat the room by creating convection currents in the air.

1 Cold air near the radiator warms up and expands.

2 This air is now warm and less dense. It moves upwards.

3 Energy is transferred between the rising warm air and the cold air surrounding it.

4 The cooler air contracts and becomes more dense. It sinks downwards.

5 Colder air moves towards the radiator.

Key fact

$$\text{density (kg/m}^3\text{)} = \frac{\text{mass (kg)}}{\text{volume (m}^3\text{)}}$$

Common error

Sometimes people describe the particles in the air as being hot or cold. However, the temperature is a property of the air, not of individual particles. However, the particles do move at different speeds, so we can talk about fast- and slow-moving particles.

Worked example

A teacher places a few crystals of a purple dye in the corner of a beaker. She heats one side of the beaker with a Bunsen burner. The purple dye moves in the pattern shown in the photo. Explain why the dye moves in this pattern.

There is an energy transfer between the Bunsen flame and the beaker. The water near the Bunsen heats up and expands. Its density decreases and so it rises upwards, taking the dye with it. It forms a layer of warmer water at the top of the beaker. As the warm water moves away from the Bunsen flame, energy is transferred between it and the surrounding water, and it starts to cool. A convection current is set up as the rising warm water is replaced by cooler water from the other side of the beaker.

Know ❯

1 What unit is used for energy transfer by convection?

2 Describe how you could measure the density of a liquid.

Apply ❯❯

3 A radiator in a room is turned on. Where is the coldest place in the room?

4 Convection currents heat up soup in a pan.

a) Use a diagram to explain how convection currents heat the soup.

b) Suggest why the soup recipe recommends stirring the soup.

Extend ❯❯❯

Fire

▲ A fire is lit at the base of one mine shaft

5 Early miners didn't have machines to help ventilate the mine shafts. Without fresh air, the miners would suffocate. To prevent this a fire would be lit at the base of one mine shaft. Explain how this would ensure all the mine had fresh air.

6 In the UK, it is often windy at the beach. This is due to convection currents. During the day, the land heats up more quickly than the sea. At night, the sea cools down more slowly than the land. Explain why it is often windy at the beach and predict which way the wind blows during the day and at night. You may find it helpful to include diagrams in your explanation.

Enquiry ❯❯❯

Convection currents in the air lead to the formation of some types of clouds.

7 Find out about the weather conditions that cause cumulus clouds, and use your knowledge of convection currents to describe how they form.

▲ Cumulus clouds are formed by convection currents in the air

» Radiation

Energy is continuously transferred from the Sun to the Earth. That energy travels through space. However, space is (mostly) a vacuum. There are almost no particles in space. So how does the Sun's energy reach us?

▲ Sunlight transfers energy from the Sun to the surface of the Earth

Key word

Radiation is a method of transferring energy as a wave.

Electromagnetic waves can travel through a vacuum. They transfer energy from the Sun to the Earth's surface. No particle movement is needed to transfer energy by **radiation**.

Infrared radiation

Infrared radiation is an electromagnetic wave which has a wavelength slightly longer than that of red light. All hot objects give out infrared radiation. Our eyes can't detect infrared radiation, but some other animals can.

Common error

Some people think that infrared radiation is similar to nuclear radiation. They aren't the same. Radiation is a term used to describe energy being transferred outwards from a central point. This happens with both types of radiation.

▲ Rattlesnakes can detect infrared radiation. They use this to help them find prey

73

▲ The heated element in the toaster emits infrared radiation to toast the bread

Grills and toasters use infrared radiation to cook food. The infrared radiation transfers energy to the outside of the food. This energy is then transferred to the inside of the food by conduction (if the food is solid) or convection (if the food is a liquid).

Seeing in the dark

Human beings and warm-blooded animals all emit infrared radiation. This infrared radiation can be seen using a special camera that can detect infrared radiation. Using this camera, even if there is no visible light, we can detect hot objects.

Infrared radiation can also pass through some materials and still be detected. This property is useful to police, firefighters and earthquake rescue teams. It allows them to find people by detecting the infrared radiation.

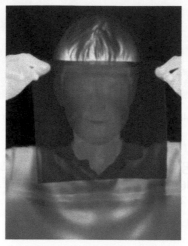

▲ Some materials allow infrared radiations to pass through them

Absorbing and emitting radiation

Different surfaces absorb infrared radiation differently. Dark colours absorb infrared radiation better than light colours, and matte surfaces absorb better than shiny surfaces. The apparatus in the diagram can be used to compare how well different surfaces absorb infrared radiation. Better absorbers heat up faster.

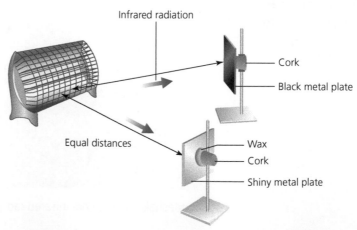

Infrared radiation

Cork

Black metal plate

Equal distances

Wax

Cork

Shiny metal plate

▶ Apparatus to investigate how well different surfaces absorb infrared radiation

White and shiny surfaces reflect infrared radiation better and so will heat up more slowly.

Dark, matte surfaces are also the best emitters of infrared radiation. Some spacecraft have dark tiles on the outside. This allows them to transfer energy quickly away from the surface as they re-enter the Earth's atmosphere and heat up.

Worked example

A thermogram is a picture taken with an infrared camera. The colours of the picture represent the temperature of the objects. Using the scale on the thermogram estimate the temperature of:

a) the person's legs
b) the sole of their trainer.

a) Between 29 ℃ (red) and 33 ℃ (pink/white).
b) The outer edges are about 20 ℃ (black/blue), but the middle does reach 30 ℃ (red).

▲ Thermogram of someone exercising on a running machine

Know >

1 How does energy reach the Earth from the Sun?

2 What is the name of the radiation emitted by the heating elements of a toaster?

Apply >>

3 The two cans shown are filled with hot water. The temperature of the water is measured for 1 hour. Predict which can will cool fastest. Explain your answer.

Shiny metal can   Matte black can

4 In hot Mediterranean countries houses are often painted white. Explain why.

5 Radiant heaters often have a shiny surface behind the heating element. Suggest the benefit of having this shiny surface.

6 A potato is being baked in the oven. Describe, as fully as you can, the energy transfers that take place to cook the potato.

Extend >>

7 Foil 'space blankets' are often used at the end of running races to make sure that the runners don't cool down too quickly. Some firefighters wear silver suits to prevent them from overheating. Explain how a shiny material can be used to both keep something warm and keep something cool.

» Extend: Forging steel

Blacksmiths work with steel. To form the steel into the shapes that they want, the blacksmiths must know the temperature of the steel and how quickly it cools down.

Traditionally, blacksmiths have used the colour of the steel to know what temperature it is. The diagram shows the colour of heated steel and the temperature when it is that colour.

	°C
White	1200
Light yellow	1100
Yellow	1050
Light orange	980
Orange	980
Light red	870
Light cherry	810
Cherry	760
Dark cherry	700
Blood red	650
Brown red	600

▲ The colour of heated steel is related to the temperature of the steel

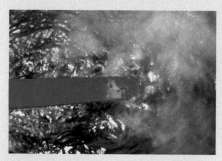

▲ Red hot metal bar being plunged into water. What is the temperature of the bar?

▲ Blacksmith's Needle, Newcastle Upon Tyne. Artwork created by the British Association of Artist Blacksmiths

Modern blacksmiths use an infrared pyrometer, which uses the infrared radiation given off by an object to measure its temperature.

The properties of steel depend on how quickly it is cooled. The faster a steel item is cooled, the harder it will be.

The graph shows two cooling curves for steel. They show the temperature change against time for two identical steel cylinders. Each cylinder was heated to 830 °C.

- One cylinder was left to cool in the air with a room temperature of 28 °C.

- The other cylinder was plunged into salt water at 24 °C.

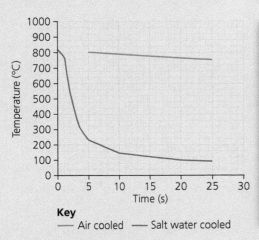

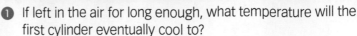

▲ Simplified cooling curves for two steel cylinders

Key
— Air cooled   — Salt water cooled

Tasks

❶ If left in the air for long enough, what temperature will the first cylinder eventually cool to?

❷ The temperature of the air was only slightly higher than the temperature of the salt water. Suggest why the cylinder plunged into the salt water cooled much more quickly than the cylinder left in the air.

❸ A third cylinder was also heated to 830°C. It was plunged into salt water at 80°C. Sketch what the cooling curve for this cylinder might look like.

▲ Buying a winter jacket. But which will keep you warmer?

Winter warmer?

If you're going skiing, it's important that you stay warm.

▲ What properties will a ski jacket need?

Clothing companies often make claims about how good their jackets are at keeping you warm.

A skiing website wants to test out the claims made about two different jackets. They want to know which jacket will keep you warmer for longer.

They send the jackets to a textile testing lab where the jackets will undergo **non-destructive testing**.

The lab uses a heated metal block and a data logger to measure the insulation properties of each coat.

Key word
Non-destructive testing is used when the object being tested must not be damaged by the test.

Tasks

❹ Plan an investigation to measure the insulation properties of each coat. You should include:
a) independent and dependent variables
b) control variables
c) number of measurements and suitable range of measurements
d) a description of how the data would be collected
e) how the data will be analysed.

Waves

Learning objectives

7 Wave effects

In this chapter you will learn...

Knowledge

- how to describe the movement of particles to and fro when some waves travel through a substance
- that energy is transferred in the direction of movement of the wave, not necessarily the particles
- that waves of higher amplitude or higher frequency transfer more energy
- the definitions of the terms ultrasound, ultraviolet (UV), microphone, loudspeaker and pressure wave

Application

- how to explain that in general electromagnetic waves cause more damage to living cells as frequency increases
- how to explain how audio equipment converts sound into a changing pattern of electric current

Extension

- how to describe how sound waves can agitate a liquid for cleaning objects, or massage muscles for physiotherapy
- how to evaluate electricity production by wave energy using data for different locations and weather conditions

8 Wave properties

In this chapter you will learn...

Knowledge

- how to describe a physical model of a transverse wave that demonstrates it moves from place to place, while the material it travels through does not
- how to describe how the properties of speed, wavelength and reflection in the model relate to transverse waves
- the definitions of the terms waves, transverse wave and transmission

Application

- how to describe the properties of different longitudinal and transverse waves
- how to use the wave model to explain observations of the reflection, absorption and transmission of a wave

Extension

- how to compare and contrast the properties of sound and light waves
- how to suggest what might happen when two waves combine

7 Wave effects

Your knowledge objectives

In this chapter you will learn:
- how to describe the movement of particles to and fro when some waves travel through a substance
- that energy is transferred in the direction of movement of the wave, not necessarily the particles
- that waves of higher amplitude or higher frequency transfer more energy
- the definitions of the terms ultrasound, ultraviolet (UV), microphone, loudspeaker and pressure wave

See page 79 for the full learning objectives.

★ **These terms are introduced and explained in Pupil's Book 1, Chapter 7.**

›› Transition: Types of wave

There isn't just one kind of wave. Physicists apply the idea of waves to lots of situations that can seem very different. Waves are a way of thinking about these situations that can help to explain how they work.

 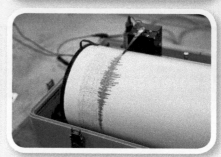

▲ Light, sound, ripples on water and earthquakes can all be described as kinds of wave

- Waves involve a **displacement** of a **medium** which is temporary.
- Something changes but after the wave it returns to the original state.
- Energy is transferred by the wave between stores. These could be the same kind of store or different kinds.
- Waves can be described in terms of their **speed**, **frequency**, **wavelength**, **time period** and **amplitude**. Not all of these are useful measurements for every kind of wave.

Absorption

The effect of **absorption** depends on what kind of wave is absorbed by what kind of material.

Sometimes the material is made to move. The energy has been transferred to a **kinetic store**. The eardrum moves when it absorbs a sound wave. Waves that cause movement are sometimes called mechanical waves.

Key words

Absorption is when some or all of the energy of a wave is transferred to a material.

A **kinetic store** of energy is filled when an object speeds up. This is also known as kinetic energy.

▲ In an earthquake, the waves can cause large movements at the surface

Sometimes the material is heated by the waves. A dark t-shirt left in the sun will become warmer as it absorbs some of the light. The energy has been transferred to a **thermal store**. (A white t-shirt reflects some of the waves so absorbs less.)

If the material absorbing the waves is made of living cells, the effects might be hard to measure. This can worry people as human senses might not be able to tell when this is happening. Some waves, for example X-rays, can damage the cells, which can go on to cause some kinds of cancer.

The energy can be transferred quickly or slowly. The effect can be large or small depending on the wave properties. For example, a bright bulb emits more light, which causes more heating in less time. The material also makes a difference; some are good absorbers, while others transmit most of the wave (and so most of the energy).

▲ The waves from some bulbs transfer energy faster than others

Worked example

Know >

1 If a wave is not absorbed by a material, what two things might happen to it?

2 What effect will a sound wave have on an eardrum?

Apply >>

3 A loud sound has a high amplitude. What words might describe a low-amplitude light wave?

4 Why can't sound travel through a vacuum? What about light?

5 Ultraviolet light is another kind of wave which can cause cells to become cancerous. Where does it come from?

Extend >>>

6 What rating is used to describe how quickly light bulbs transfer energy? Why might an old-fashioned filament bulb and a modern LED version have different numbers but be of equal brightness?

Key fact

There are many other kinds of damage which can also lead to cancer for example chemical (tar in cigarette smoke causes lung cancer) and viral (HPV is a virus which can cause cervical cancer).

» Core: Sounds and explosions

Sound travels from a vibrating source to a detector through a medium. The particles in the medium are made to vibrate in a particular pattern, moving back and forth so the energy is transferred. Anything which is made to move by the vibrating medium can be used as a detector.

Sound in a room can be detected by measuring the vibration of windows with a laser. Spies can use this to eavesdrop on conversations in distant buildings. An easy way to model this is to hold a blown-up balloon close to a loudspeaker, or a piece of paper in front of the face of a talking person. The movement of the balloon or paper can be detected by the person holding the object.

Anybody who's stood close to a big loudspeaker knows the sound is not just heard. High-amplitude sounds, especially at low frequencies, can be felt through the feet or in the chest. This is because sound waves in the air are just a particular kind of **pressure wave**. The particles are compressed as the wave passes.

An explosion also causes a pressure wave. This might be detected as a sound or a physical impact, depending on how big the explosion and how close it is. The sudden high pressure can cause a great deal of damage. Usually an explosion is a single wave pulse, like one ripple in water.

Sound and current

Sound waves can be converted into electrical signals and back using devices like a **microphone** and a **loudspeaker**. This shows the link between amplitude of the wave and loudness, the pitch of the sound and frequency, and so on. The line that is drawn shows the displacement on the *y*-axis with time on the *x*-axis.

▲ The sound from a large speaker can be felt as well as heard

Key word

A **pressure wave** like sound has repeating patterns of high-pressure and low-pressure regions.

Key words

A **microphone** converts a sound wave into an electrical signal. A **loudspeaker** converts an electrical signal into a sound wave.

(a) 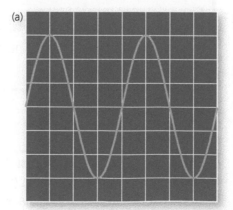 (b)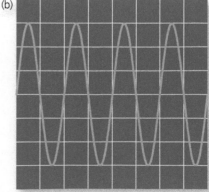

▲ The closer together the waves, the higher the frequency

Because sound is a longitudinal wave, measuring the wavelength is quite hard. An equivalent wave in a slinky can be measured by filming the wave, but sound waves are invisible. The speed can be measured more easily, either by timing an echo or measuring the time lag between the flash and bang of an exploding balloon across the school field.

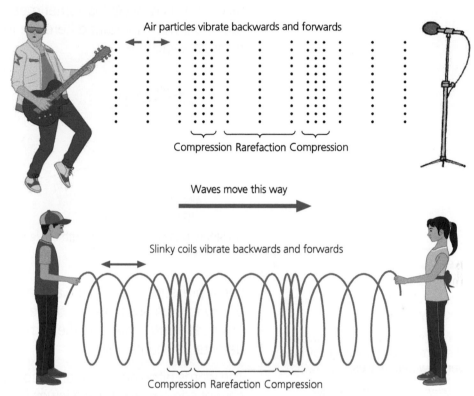

Air particles vibrate backwards and forwards

Compression Rarefaction Compression

Waves move this way

Slinky coils vibrate backwards and forwards

Compression Rarefaction Compression

▲ Wavelength is not always easy to measure

Know >

1 A sound wave and a shock wave from an explosion both cause movement in the absorber. What would the difference be?

Apply »

2 What device turns sound waves into an electrical signal? Give an example of one being used.

3 In the picture above, the technician is cleaning the engine part without using soap, bleach or anything else that might contaminate it. Explain how ultrasonic cleaning works.

Extend »»

4 Describe the measurements and calculations needed to find the speed of sound in air by either method mentioned.

5 Explain how a very loud sound could damage the eardrum, using the idea of absorption.

Key fact

Sound, especially high-frequency sound, can be used to make objects vibrate to shake off dirt or to massage aching muscles.

» Core: Nearly visible

Light and sound both act like waves but are otherwise quite different. Sound is a pressure wave which needs particles as a medium. The displacement of light, instead of particles being pushed or pulled, involves changes to magnetic and electric fields. This is why light is sometimes called electromagnetic or EM radiation. The important difference for now is that light can travel through a vacuum.

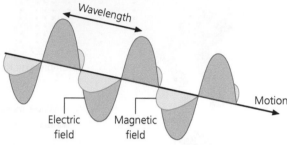

▲ Light waves change the strength of electric and magnetic fields, not the position of particles

Key words

Ultrasound waves have frequencies higher than the human auditory range.

Infrasound waves are outside of the human auditory range, by having very low frequencies.

Ultraviolet (UV) waves have frequencies higher than visible light.

Infrared (IR) waves have frequencies lower than visible light.

Some pressure waves in the air have a frequency which is too high or low for humans to hear – these are called **ultrasound waves** or **infrasound waves**. In the same way, not all light waves can be detected by human eyes.

The words ultra- and infra- are also used for kinds of light which we cannot see. **Infrared** has a lower frequency than (red) visible light and is what scientists mean by 'thermal radiation'. **Ultraviolet** is important because it can cause damage to living cells. In different amounts it causes tanning, sunburn and skin cancer. Human eyes cannot detect either ultraviolet or infrared waves.

▲ Not all light waves are visible to human eyes

Damaging cells

Scientists use the idea of waves to explain some very different situations. This means people can get confused about the differences between safe and dangerous waves. When UV waves are absorbed by living cells, they can cause damage. The longer a person is exposed, or the stronger the UV, the more the cells are damaged.

Visible light can't cause the same kind of damage, no matter how long it shines on a person's skin. Sunlight is a mixture of different frequencies of light – this is why a rainbow can be made by splitting up sunlight into different colours – and UV is one of them. Scientists have collected data on rates of skin cancer to learn more about what makes the UV part of sunlight dangerous.

UV can damage other materials too. It all depends on how the waves are absorbed and the effect this has on the chemical bonds in the material.

▲ Sunburn is a sign of damage to cells in the skin

Know >

1 What kind of light does sunblock absorb?

Apply >

2 Is a Mexican wave or a shoulder-barge a better model for sound waves? A diagram may help to explain your answer.
3 A tourist at the beach is surprised by their sunburn because the breeze meant they didn't feel hot all day. Explain their mistake.
4 Which frequencies of EM radiation are dangerous to cells, higher or lower than visible light?

Extend >

5 Research the recommended daily guidelines for sunlight exposure in the UK. Explain how they balance the risks and benefits.

Key facts

Just as ultrasound is based on the *human* auditory range, infrared and ultraviolet are visible for some animals.

Too much sunlight exposure can cause damage to skin cells. But too little means the skin doesn't make vitamin D, which can lead to the deficiency disease rickets.

» Extend: Water waves

Ripples on water are waves. For many non-scientists they are the first example that come to mind. Like sound waves they are caused by a moving object and cause movement in turn. The distance between wave peaks is called the **wavelength**.

▲ Waves at the beach arrive at a regular time interval

Because the displacement of a sound wave is in the *same* direction as the wave travels, it is hard to measure the wavelength. Sound is a **longitudinal wave**, with displacement in the *same* direction as the wave travels. This is not true for water waves, where the displacement is at a *right angle* to the wave travel. They are called **transverse waves**.

Key words

The **wavelength** is the length of one complete wave pattern. It is measured in metres (m) and has the symbol λ.

The displacement of a **longitudinal wave** is along or in line with the direction of wave travel.

Transverse waves like light and ripples on water have displacement which is perpendicular or at a right angle to the direction of wave movement.

▲ As the wave moves along, the water moves up and down

Anything that is floating in the water moves up and down with the medium. It will only be moved in the direction of the wave if it tilts, like a surfboard. Floating bottles can only carry messages from a desert island if they are carried by the current, rather than waves. But the movement of a float can be used to generate electricity.

Wave power

A hydroelectric power station uses water running downhill to turn a turbine. In a tidal barrage, the water moves because of the tides. Wave power uses the movement of a float up and down as the waves move along the surface off the sea.

Like any other method of generating electricity, the benefits and problems need to be compared. Wave power produces no carbon dioxide and has a low running cost as there is no fuel. There are concerns about how it may affect fish and other marine life.

Instead of just affecting the beach and coastline (usually by causing erosion), there is also an effect on the floats. A small proportion of the energy is absorbed during this process, transferred to the kinetic store of the floats.

Common error

Students often use the words 'water power' when they need to specify which of hydroelectric, tidal and wave they mean.

Tasks

1. What is the unit of wavelength?
2. Why might fishing crews object to generating electricity by wave power?
3. How would you increase the amplitude of a water wave?
4. A prototype wave power float produces different amounts of electricity depending on the weather. The main factor affecting this is the wind.

Wind speed (mph)	Electricity generated in 24 hours (kWh)
0–15	160
16–30	400
31–45	720
46–60	480

   a) Which range of wind speeds produces the most electricity?
   b) In a week there are two calm days (less than 15 mph), two breezy days (16–30 mph), one windy day (31-45 mph) and two stormy days (46–60 mph). How much electricity is generated during the week?
   c) During one day the wind speed varies from 17 to 40 mph. Estimate the average electricity generated.
5. Describe how you could measure the frequency and wavelength of water waves at the beach. What equipment would you need?

8 Wave properties

» Transition: Reflection

The idea of a wave is a model that scientists use to describe several situations that seem very different. If the same model can be used, this means there must be similarities as well as differences. Waves all behave in particular, predictable ways in the right circumstances.

All waves can be reflected by some surfaces. For visible light, the surface needs to be shiny and polished. Even materials like water and glass, which seem transparent, will reflect some light from the surface.

▲ The image of the trees and buildings is reflected in the water

All waves reflect – but from different surfaces. Echoes are the **reflection** of a sound, and work best with a smooth, hard surface. This is something every small child demonstrates when they go into a cave or tunnel.

Your knowledge objectives

In this chapter you will learn:
• how to describe a physical model of a transverse wave that demonstrates it moves from place to place, while the material it travels through does not
• how to describe how the properties of speed, wavelength and reflection in the model relate to transverse waves
• the definitions of the terms waves, transverse wave and transmission

See page 79 for the full learning objectives.

Key words

A **wave** always involves the transfer of energy by vibrations from one place or material to another, without matter being permanently moved.

Reflection is when a wave changes direction away from a surface.

► Sound reflecting against the sides of a room or other space is called an echo

Absorption and transmission

Most materials are not perfect reflectors, so there is some **absorption**. This means that there is a transfer of energy to the material. The effect will depend on the kind of wave and the kind of material (see Pupil's Book 1, Chapter 7 for details).

In many cases there is *some* reflection, *some* absorption and *some* transmission. If the wave can travel through the material then it has been transmitted. Scientists use simple models, often assuming only one or two of these three things happens.

Thinking about light and a tinted surface may help to explain these ideas. Some of the light is reflected. Some is absorbed; it actually causes a tiny amount of heating. Some is transmitted, although it may change direction at the surface because of **refraction**.

> ### Key words
>
> **Transmission** is when the wave goes through the material instead of being absorbed or reflected.
>
> **Refraction** is a change in the direction of light when it goes from one material into another. The new direction, measured from the normal, is the angle of refraction.

▲ The incident, reflected and transmitted rays all can be seen here

Know >

1 What three things can happen when a wave meets a boundary between materials?

2 When a teacher speaks in a room, what is vibrating as the sound wave travels?

3 What word is used for a reflected sound wave?

Apply >>

4 If a material absorbs some of the energy from light is it translucent or transparent?

5 Sketch the glass block and light rays from the diagram above and label each of the rays.

» Core: Numbers

The behaviour of different waves is similar. Another way to understand the links between ripples in water, sound in air and light in a vacuum is to consider the measurements that scientists make. Although the numbers are different – in some cases by factors of millions – the variables are the same.

Type of wave	Typical speed of wave
Ripples on water	10 m/s
Sound in air	330 m/s
Light in vacuum	300 000 000 m/s

In all cases, something is displaced from the original position. The size of that displacement is the **amplitude** and is linked to the energy transferred by the wave. With sound it describes the temporary movement of particles.

One of the easiest examples to observe in the school science lab uses a spring or slinky laid on a table. Instead of pushing or pulling the end, it is flicked to the side to cause a **transverse wave**.

This makes it much easier to see and measure the **wavelength**; this is the distance between two peaks. The amplitude is a measurement of how far the spring moves (sideways) from the middle.

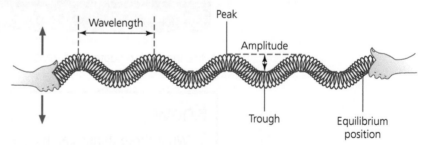

▲ A slinky wave can be measured with a ruler and stopwatch

The speed of a wave can be measured too – simply measure how far a peak travels in each second. In most cases the tricky part is keeping track of which wave is being timed. This is much easier if the movement can be recorded and played back.

Worked example

Measuring wavelength is straightforward. Wave speed is calculated in exactly the same way as the speed of any other moving object. The distance travelled in metres is divided by the time taken in seconds.

A ripple in a swimming pool travels 8 m in 4 s. What is its speed?

$$speed\ (m/s) = distance\ (m)\ /\ time\ (s)$$

$$v = d\ /\ t$$

$$= 8\,m\ /\ 4\,s$$

$$= 2\,m/s$$

Know >

1 A high-amplitude sound wave is *loud*. What word might be used to describe a high-amplitude light wave?

Apply »

2 A wave travels from one end of a slinky to the other in 1.5 s. It is 2 m long. What is the speed?

3 A sound wave takes 0.5 s to travel from a speaker to a microphone on a school field. What is the distance travelled?

4 The Earth is 150 million km from the Sun. How long does it take light to travel from the Sun to our planet's surface?

Extend »»

5 Earthquakes also involve waves, both longitudinal and transverse. Briefly research the cause and medium of these seismic waves, writing a paragraph about how the amplitude is measured.

Key facts

For some waves, like light, the amplitude is a more complicated measurement. It always links to the energy put in by the source.

A Mexican wave is also a transverse wave – people's arms are moving at a right angle to the direction of wave travel.

Common error

The period of a wave is how much *time* it takes one wavelength to pass a point. The wavelength is the *distance* between one wave and the same point on the next wave. 'How long' can mean either.

» Core: Waves and time

The frequency of a wave is a count of how many waves pass a point each second. If the time taken for a single wave to pass that point (called the period) is more than a second, the frequency will be less than one. Frequency is measured in Hertz (Hz).

These measurements can't be made from a static picture. Instead, like speed, observers must use a stopwatch and count. As with other situations, it is much easier if the wave can be filmed in motion, whether ripples on water or a wave on a slinky.

▲ Buses outside the Houses of
Parliament, London

▲ From this viewpoint an observer could record how many waves arrive in a
given time

A good way to think about what frequency and period mean
uses a bus timetable. If three buses go past a stop each hour, the
frequency is three buses per hour. The time between buses must be
20 minutes – or to make the link clear, one-third of an hour.

It is worth remembering that these numbers don't reveal anything
about the speed of the bus. For that, another number is needed –
the distance travelled during the bus journey.

Two waves at once

So far, all the examples have used waves travelling in the same
direction and at the same speed. This is not what always happens.
Consider a paddling pool and a child with two toys. They are dropped
in, first one then the other, to see the effects. Each sends a wave
across the surface, which reflects from the far side and travels back.

▲ When children play with water
making waves is part of the fun

There is a point when two waves are travelling in opposite
directions. Each wave is causing displacement, pushing the
water up as the wave travels along. When two waves meet, they
might push the water up by more than either would alone; the
displacements add together.

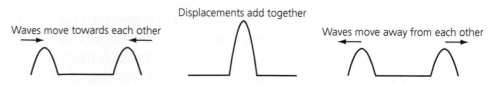

Displacements add together

Waves move towards each other   Waves move away from each other

▲ Sometimes the effects of the waves add up

If the waves that meet have displacements in opposite directions, the
effect is to cancel out instead of add together. This can be seen when
a transverse wave is sent along a slinky to reflect from a fixed end. The

reflection has a displacement in the opposite side. The two waves are trying to push the slinky in opposite directions, so nothing happens.

Waves move towards each other   Displacements cancel out   Waves move away from each other

 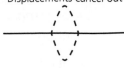 

▲ Sometimes the effects of the waves cancel out

This effect is called superposition, which means the two waves are overlapping. The effect can be seen when two sources of the same kind of wave affect each other. Two loudspeakers in a room will cause loud and quiet spots if not aimed correctly.

Worked example

Frequency describes how often waves arrive (i.e. how frequent they are) in waves per second or Hertz (Hz). The period is the time taken for one complete wave to pass. The link between then is simple:

$$f = \frac{1}{T} \quad \text{or} \quad T = \frac{1}{f}$$

where f = frequency (Hz) and T = time period (s)

If it takes 4 s for a wave to pass a given point, then

$$f = \frac{1}{T}$$

$$= \frac{1}{4}$$

$$= 0.25\,\text{Hz}$$

Know >

1 What happens to the time period if the frequency increases?

Apply >>

2 What is the frequency for a wave with a time period of 0.1 s?

3 What is the time period for a wave with a frequency of 50 Hz?

4 There are five water ripples each minute. What is the frequency?

Extend >>>

5 Investigate the science involved in the design of noise-cancelling headphones; summarise the link to superposition.

Common error

t is used for any measurement of time while *T* is the time taken for one complete wave to pass, but they have the same unit of seconds (s).

Key fact

The frequency and period of a wave are opposites. As one increases the other decreases.

» Extend: Wave equation

Studying wave behaviour shows that waves travel:

- by *temporarily* displacing a medium and *permanently* transferring energy

- at a uniform (constant) speed if the medium is consistent

- in straight lines until the medium changes, when they are reflected, absorbed or transmitted.

The quantities of frequency, wavelength and speed help to describe this movement in numbers. They are linked together by an equation which works for *any* wave:

wave speed (m/s) = frequency (Hz) × wavelength (m)

$v = f \times \lambda$

A good way to imagine what is happening uses a train travelling on a track. Each carriage is the same length, 8 m. Four carriages pass by an observer in a second. This is enough information to work out the speed: the train must have moved 4 × 8 = 32 m in each second, so the speed is 32 m/s.

▲ Each carriage is the same size

Train model	Wave
Length of carriage	Wavelength
Carriages per second	Frequency
Train speed	Wave speed

Worked example

Water waves in a harbour are 15 m apart. The frequency is 0.5 Hz. What is the speed?

$v = f \times \lambda$

$= 0.5 \times 15$

$= 7.5 \text{ m/s}$

Common error

Remember to convert any values into standard units, e.g. 1 kHz = 1000 Hz.

The equation can be rearranged to make frequency or wavelength the subject:

$f = v/\lambda$ or $\lambda = v/f$

Tasks

1. A vibration in a steel spring travels as a wave. The wavelength is 2 m and the frequency is 3 kHz. What is the speed?
2. The speed of sound in air is 330 m/s. What is the wavelength of a sound with a frequency of 400 Hz?
3. A seismic wave travels through the Earth's crust at 4 km/s. If the wavelength is 80 km, what is the frequency in Hertz?

Light waves

Each wavelength of visible light is a different colour. Longer wavelengths are towards the red end of the spectrum, while blue and purple have a shorter wavelength. This works in the same way as high- and low-pitched sounds. The numbers, however, are very different to anything people normally think about measuring.

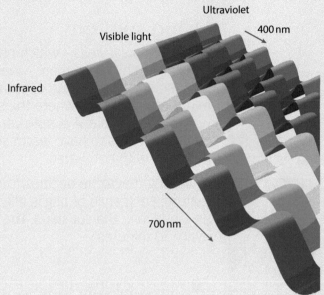

Ultraviolet

400 nm

Visible light

Infrared

700 nm

▲ Each wavelength is a different colour

Key facts

The speed of light is 3×10^8 m/s. This is 300 000 000 m/s.
1 nanometre (nm) is one thousand-millionth of a metre or 1×10^{-9} m.
Standard form is used to show very large and very small numbers. A calculator will use a button marked EXP or similar. If the number after the ten is positive it is a large number. If it is negative, it is a small number. For example, $400\,000 = 4 \times 10^5$ and $0.003 = 3 \times 10^{-3}$.

▲ Can you find the same functions on your calculator?

Light waves follow the same equation as all other waves. Instead of *v* for the wave speed, the letter *c* is used because light always travels at the same speed in a vacuum. According to some sources, *c* was originally short for *celeritas*, which means fast. And it is very, very fast.

Worked example

Red light has a wavelength of 700 nm. What is the frequency?

$$c = f \times \lambda$$

$$f = \frac{c}{\lambda}$$

$$= \frac{3 \times 10^8}{7 \times 10^{-7}}$$

$$= 4.3 \times 10^{14} \text{ Hz}$$

Tasks

4. Violet light has a frequency of 7.5×10^{14} Hz. Show that the wavelength is approximately 400 nm.
5. The human hearing range for sound is 20 to 20 000 Hz. Use the information above to give the human visual range for light.

Enquiry:
Solar showers and insulated mugs

▲ (a) A solar *cell* produces a current.
(b) A solar *panel* heats up water or other liquids

▲ The water is heated by the sunlight

Using sunlight

As well as solar *cells* – which produce a potential difference when light shines on them – we use sunlight to heat things up in solar *panels*. These approaches are both useful when in remote places, or where the supply of electricity or hot water is unreliable. A simple version of a solar panel is used by backpackers to have warm showers even when miles from buildings with hot water.

A walking magazine decides to test some of these products for a feature. Each style of bag is filled with water and placed outside on a bench in the sunshine. The feature writer tested the water every 30 minutes.

❓

❶ Why is it important they are all outside in the same place?
❷ Fill in the blank for the question they were investigating: How does … change over time?
❸ What measuring device would be used? What are the units?

After reading the feature, a new company think they can produce better solar showers. They start by getting samples of different materials to use for the bags. One designer is looking at which colour is best. They do this by wrapping a water bottle in each sample material, leaving under a desk light for a fixed amount of time and then measuring the increase in temperature.

❓

❹ Should the designer leave the wrapped bottles for a short or long period of time? Why?

The shape of the bag will also be important. They decide to keep the volume constant but change how deep the water inside will be when laid flat. The first two designs give almost the same result. The temperature increase of water in the 1 cm deep bag is 8 °C compared to 9 °C for the bag which is 5 cm deep. The designers decide to collect more data and make prototypes that are 2, 3 and 4 cm deep.

The designers decide to collect more data and make prototypes with depths of 2, 3 and 4 cm.

⑤ Why is it better to do more tests?

The new results are added to a new table.

Depth of water (cm)	Temperature increase (°C)
1	9
2	13
3	12
4	10
5	8

⑥ Which prototype should be used for the new product? Why?
⑦ Use the word 'conduction' to explain why the thinnest prototype might *lose* heat faster than the others.

Quality control

The designers are asked to investigate problems with another product by the company. Their thermal mugs are supposed to keep drinks warm for hours but some users complain that they don't work very well. The designers pour hot water from the kettle into three identical mugs. They place a temperature probe in each and seal the lids. The mugs are placed in ice water to simulate being outside in winter.

⑧ Identify the main risks and hazards for this test.
⑨ What sensible precautions could they take to avoid injury?
⑩ What approximate outside temperature is being simulated?
⑪ Draw a results table for the experiment assuming measurements every 10 minutes for 2 hours. Include units in the headings of each of the five columns.

▲ The mug is designed to keep a drink warm on a campsite

Matter

Learning objectives

9 Periodic table

In this chapter you will learn...

Knowledge

- that metals are generally found on the left side of the table, and non-metals on the right
- that the elements in a group typically react in a similar way and sometimes show a pattern in reactivity
- that as you go down a group and across a period the elements show patterns in physical properties
- the names of groups 1, 7 and 0 and give general descriptions of the characteristic properties of the elements in each of these groups
- the definitions of the terms periodic table, physical properties, chemical properties, groups and periods

Application

- how to describe a trend in physical properties when given data
- how to describe the reactions of an unfamiliar group 1 or 7 element
- how to use data showing a pattern in physical properties to estimate a missing value for an element
- how to use observations of a pattern in chemical reactions to predict the behaviour of an element in a group

Extension

- how to predict the position of an element in the periodic table based on information about its physical and chemical properties
- how to choose elements for different uses from their position in the periodic table
- how to use data about the properties of elements to find similarities, patterns and anomalies

10 Elements

In this chapter you will learn...

Knowledge

- that most substances are not pure elements, but compounds or mixtures, which have different properties to the elements they contain
- how to name simple compounds
- the symbols of some common elements
- the definitions of the terms element, atom, molecule, compound, chemical formula and polymer

Application

- how to interpret a chemical formula to identify the elements present, their relative proportions and suggest its name
- how to interpret particle diagrams which relate to elements, mixtures or compounds and molecules or atoms
- how to use observations from chemical reactions to decide if an unknown substance is an element or a compound

Extension

- how to use particle diagrams to predict physical properties of elements and compounds
- how to deduce a pattern in the formula of similar compounds and use it to suggest formulae for unfamiliar ones
- how to compare and contrast the properties of elements and compounds and give a reason for their differences
- how to describe and explain the properties of ceramics and composites

9 Periodic table

» Transition: The periodic table

There are approximately 120 chemical **elements**. Elements are listed on the **periodic table** in a way that allows us to understand patterns in their properties and reactions.

Your knowledge objectives

In this chapter you will learn:
- that metals are generally found on the left side of the table, and non-metals on the right
- that the elements in a group typically react in a similar way and sometimes show a pattern in reactivity
- that as you go down a group and across a period the elements show patterns in physical properties
- the names of groups 1, 7 and 0 and give general descriptions of the characteristic properties of the elements in each of these groups
- the definitions of the terms periodic table, physical properties, chemical properties, groups and periods

See page 99 for the full learning objectives.

Key words

An **element** is a pure substance made from only one type of atom. Elements are listed on the periodic table.

The **periodic table** lists all of the chemical elements in groups and periods.

A **group** on the periodic table is a vertical column.

Columns and rows

Each vertical column in the periodic table is called a **group**. Each group contains elements which are similar to each other in their properties and chemical reactions.

1	2											3	4	5	6	7	8 or 0
						H Hydrogen											He Helium
Li Lithium	Be Beryllium											B Boron	C Carbon	N Nitrogen	O Oxygen	F Fluorine	Ne Neon
Na Sodium	Mg Magnesium											Al Aluminum	Si Silicon	P Phosphorus	S Sulfur	Cl Chlorine	Ar Argon
K Potassium	Ca Calcium	Sc Scandium	Ti Titanium	V Vanadium	Cr Chromium	Mn Manganese	Fe Iron	Co Cobalt	Ni Nickel	Cu Copper	Zn Zinc	Ga Gallium	Ge Germanium	As Arsenic	Se Selenium	Br Bromine	Kr Krypton
Rb Rubidium	Sr	Y	Zr	Nb	Mo	Tc	Ru	Rh	Pd	Ag Silver	Cd	In	Sn	Sb	Te	I Iodine	Xe Xenon
Cs Caesium	Ba	La	Hf	Ta	W Tungsten	Re	Os	Ir	Pt Platinum	Au Gold	Hg Mercury	Tl	Pb	Bi	Po	At Astatine	Rn Radon
Fr	Ra	Ac	Rf	Db	Sg	Bh	Hs	Mt	Ds	Rg	Cn	Nh	Fl	Mc	Lv	Ts	Og

▲ Groups have been labelled in this periodic table. Group 1 has been shaded in blue. Group 7 has been shaded in green. Group 8 is also called Group 0 and has been shaded in yellow

Each row in the periodic table is called a **period**.

Key word

A **period** on the periodic table is a horizontal row.

1	H Hydrogen																	He Helium
2	Li Lithium	Be Beryllium											B Boron	C Carbon	N Nitrogen	O Oxygen	F Fluorine	Ne Neon
3	Na Sodium	Mg Magnesium											Al Aluminum	Si Silicon	P Phosphorus	S Sulfur	Cl Chlorine	Ar Argon
4	K Potassium	Ca Calcium	Sc Scandium	Ti Titanium	V Vanadium	Cr Chromium	Mn Manganese	Fe Iron	Co Cobalt	Ni Nickel	Cu Copper	Zn Zinc	Ga Gallium	Ge Germanium	As Arsenic	Se Selenium	Br Bromine	Kr Krypton
5	Rb Rubidium	Sr	Y	Zr	Nb	Mo	Tc	Ru	Rh	Pd	Ag Silver	Cd	In	Sn	Sb	Te	I Iodine	Xe Xenon
6	Cs Caesium	Ba	La	Hf	Ta	W Tungsten	Re	Os	Ir	Pt Platinum	Au Gold	Hg Mercury	Tl	Pb	Bi	Po	At Astatine	Rn Radon
7	Fr	Ra	Ac	Rf	Db	Sg	Bh	Hs	Mt	Ds	Rg	Cn	Nh	Fl	Mc	Lv	Ts	Og

▲ Periods have been labelled on this periodic table. Period 1 only contains two elements and is shaded in purple. Period 2 is shaded in brown and Period 3 is shaded in grey

Each element can be given a unique 'address' on the periodic table using its group number and period number. This is like a set of coordinates in maths. For example, sodium is the element in Group 1 and Period 3. Nitrogen is in Group 5 and Period 2.

Worked example

Which element is in Group 7 and Period 5?

Iodine is in Group 7 and Period 5.

Apply »

1 What name is given to a vertical column in the periodic table?

2 What name is given to a horizontal row in the periodic table?

3 Name two elements in Group 1.

4 Name two elements in Period 2.

» Core: The structure of the periodic table

All of the chemical elements are listed on the periodic table. However, the elements are not listed in alphabetical order. They are arranged in a way that allows us to spot patterns in their properties and the way that they react. Metals are found on the left and in the middle of the periodic table. Non-metals are found on the right.

Groups and periods

The chemical elements were not added to the table in the order they were discovered. The arrangement of the elements in the periodic table means that elements with similar chemical properties are arranged in the same vertical column. The columns in the periodic table are called groups. As you go down a group, there is often a trend in properties like reactivity and melting point. Some of the groups have specific names.

As you go across a period, the way a property like melting point changes is usually similar to the pattern in the period above.

	1	2											3	4	5	6	7	8 or 0
1							**H** Hydrogen											**He** Helium
2	**Li** Lithium	**Be** Beryllium											**B** Boron	**C** Carbon	**N** Nitrogen	**O** Oxygen	**F** Fluorine	**Ne** Neon
3	**Na** Sodium	**Mg** Magnesium											**Al** Aluminum	**Si** Silicon	**P** Phosphorus	**S** Sulfur	**Cl** Chlorine	**Ar** Argon
4	**K** Potassium	**Ca** Calcium	**Sc** Scandium	**Ti** Titanium	**V** Vanadium	**Cr** Chromium	**Mn** Manganese	**Fe** Iron	**Co** Cobalt	**Ni** Nickel	**Cu** Copper	**Zn** Zinc	**Ga** Gallium	**Ge** Germanium	**As** Arsenic	**Se** Selenium	**Br** Bromine	**Kr** Krypton
5	**Rb** Rubidium	**Sr**	**Y**	**Zr**	**Nb**	**Mo**	**Tc**	**Ru**	**Rh**	**Pd**	**Ag** Silver	**Cd**	**In**	**Sn**	**Sb**	**Te**	**I** Iodine	**Xe** Xenon
6	**Cs** Caesium	**Ba**	**La**	**Hf**	**Ta**	**W** Tungsten	**Re**	**Os**	**Ir**	**Pt** Platinum	**Au** Gold	**Hg** Mercury	**Tl**	**Pb**	**Bi**	**Po**	**At** Astatine	**Rn** Radon
7	**Fr**	**Ra**	**Ac**	**Rf**	**Db**	**Sg**	**Bh**	**Hs**	**Mt**	**Ds**	**Rg**	**Cn**	**Nh**	**Fl**	**Mc**	**Lv**	**Ts**	**Og**

▲ In this periodic table, the groups and periods have been numbered. Group 1 is shaded in blue, Group 7 is shaded in green, and Group 8 (also called Group 0) is shaded in yellow. The elements in Period 2 have a thick black box around them. The transition metals are shaded in purple

As we have seen, each element has a unique 'address' in the periodic table. Boron is in Group 3 and Period 2. Potassium is in Group 1 and Period 4. Be careful that you don't get the group and period confused!

The transition metals

In the middle of the periodic table is an area which isn't usually split into groups. It contains the elements called the transition metals. Most of the common metals with typical properties are found here.

The noble gases

Key word

A chemical is **inert** if it doesn't react with other chemicals.

The elements in Group 0 (also called Group 8) are called the noble gases. These gases are all **inert**, which is another word for unreactive. The elements in Group 0 do not normally react with other chemicals and do not normally form compounds. But that actually makes them quite useful!

Here are some uses for some of the elements in Group 0.

Element	Used for...	Because...
Helium	Party balloons and airships	It has a very low density (lighter than air).
Neon	Glowing advertising signs	It gives off a coloured light when electricity passes through it.
Argon	Welding metals	It is unreactive, so used in welding to prevent the hot metal reacting with oxygen in the air.
Krypton	High-power lights	It gives off a bright light when electricity passes through it.

▲ So-called neon signs are filled with either neon, argon or krypton

▲ Helium is commonly used for party balloons

As you go down the elements in Group 0, there is a trend in their densities. Helium is the least dense – much lighter than air. Neon is a little bit more dense, but still lighter than air. Argon is slightly more dense than air. The noble gases get more dense as you go down Group 0.

Worked example

Which is the noble gas with the third lowest density? Explain how you worked out your answer.

Argon is the noble gas with the third lowest density. The density of the noble gases increases as you go down the group, so the third element going down Group 0 is argon.

Know >

1 What name is given to the elements in Group 0?

2 Give one use for argon.

3 The noble gases are inert. What does this mean?

Apply >>

4 Which element is in Group 1 and Period 2?

5 Krypton and neon gases give off light when an electrical current is passed through them. Would you expect helium to do the same? Explain your answer.

Extend >>>

6 Before aeroplanes were invented, airships were used to transport passengers. Initially, they used hydrogen as the gas in the balloon which caused the airship to float. A number of deadly accidents caused the hydrogen airship to be replaced by helium airships. Use your knowledge of the properties of hydrogen and helium to suggest what happened in the accidents and why helium is a safer gas to use.

Enquiry >>>>

7 Use secondary sources to find out why hydrogen is not placed in a group in the periodic table.

>> Core: Group 1 – the alkali metals

The elements in Group 1 are called the alkali metals. They are all very reactive and produce alkaline solutions when they react with water. There are trends in **chemical properties** as you go down the group.

Reactions with oxygen

The alkali metals must be stored in bottles which are full of oil because they react quickly with oxygen in the air. To begin with, the surface of the metal looks dull grey, but when they are cut with a sharp knife, the metal shows its true shiny silver colour. As you go down the group, there is a trend (a pattern) in the reactivity of the metals with oxygen.

Key word

Chemical properties are a description of the way that a substance reacts with other chemicals. Being reactive is a chemical property. So is being acidic.

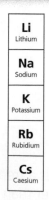

▲ The alkali metals

Metal	Reaction with air
Lithium	The shiny surface turns dull within 1 minute
Sodium	The shiny surface turns dull within 10 seconds
Potassium	The shiny surface turns dull almost immediately

Reactions with water

The alkali metals are very unusual because they float on water. No other metals do this. Hopefully your teacher will demonstrate these reactions to your class. There is a trend in reactivity – as you go down the group, the metals get more reactive.

▲ The reaction of potassium with water is very exciting!

Metal	Reaction with water
Lithium	Floats, fizzes, moves around slowly. The solution left behind has a pH of 14.
Sodium	Melts into a ball, floats, fizzes and moves quickly. The pH of the solution produced is 14.
Potassium	Floats and sparks. Gas is produced which burns with a purple flame. The pH of the solution is 14.

The gas being produced in each case is hydrogen. The solution produced is alkaline because a metal hydroxide is made. Here is an example of a word and symbol equation for the reaction:

sodium + water → sodium hydroxide + hydrogen

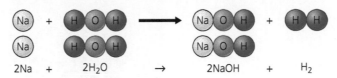

$$2Na + 2H_2O \rightarrow 2NaOH + H_2$$

Worked example

Write a word equation for the reaction between lithium and water.

lithium + water → lithium + hydrogen
                           hydroxide

Know >

1 What is the name of the elements in Group 1?

2 Why do the metals in Group 1 float on water?

3 Why are the metals in Group 1 stored in oil?

Apply >>

4 Write a word equation for the reaction between potassium and water.

5 Predict how long it would take for rubidium to react with oxygen when it is freshly cut with a sharp knife.

6 Name the products of the reaction of caesium and water.

Extend >>>

7 Write a symbol equation for the reaction of lithium with water.

8 Sodium floats on water. Suggest why it would not be a sensible metal to use for making a boat.

Enquiry >>>>

9 Using the information on these pages, suggest what safety precautions should be taken when demonstrating the reaction of rubidium and caesium with water.

Key word

Physical properties are a description of the way that a substance behaves which don't involve it reacting with other chemicals. Melting point, electrical conductivity and hardness are examples of physical properties.

» Core: Group 7 – the halogens

The non-metal elements in Group 7 are called the halogens. The name halogen means 'salt maker' because halogens react with metals to make salts. Common salt is sodium chloride, and is made when sodium reacts with chlorine. There are trends in chemical properties and **physical properties** as you go down the elements in Group 7.

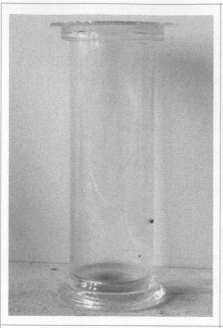

Chlorine

Bromine

Iodine

| F |
| Fluorine |
| **Cl** |
| Chlorine |
| **Br** |
| Bromine |
| **I** |
| Iodine |
| **At** |
| Astatine |

▲ The halogens

Trend in reactivity

The halogens are all reactive, but the most reactive of all of the elements on the periodic table is fluorine, right at the top of Group 7. Fluorine is so reactive that it even reacts with some noble gases! It is very hard to store fluorine safely because you need to use a special container which the fluorine won't react with.

Trend in boiling point

Looking at the images above, you can see that chlorine is a green gas, bromine is a dark brown liquid and iodine is a grey solid. This tells us that it is easier to turn chlorine into a gas than the other two – it has a lower boiling point. To turn bromine into a gas only requires a little bit of gentle heating, but iodine is a solid and so it has a much higher melting point. We can show the boiling points on a graph.

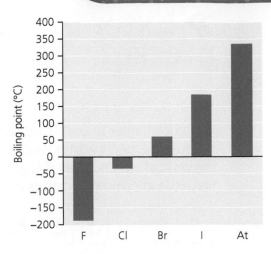

▲ A bar chart showing the boiling points of the halogens

Worked example

Suggest whether astatine will be a solid, liquid or gas. Explain your answer.

Astatine will be a solid. As you go down Group 7, the boiling points and melting points increase. Because iodine is a solid, astatine must also be a solid.

Know >

1 What is the name given to the elements in Group 7?

2 Which is the most reactive element in Group 7?

3 Are the elements in Group 7 metals or non-metals?

Apply »

4 Suggest whether fluorine is a solid, liquid or gas. Explain your answer.

5 Predict the chemical reactivity of iodine, relative to the other halogen elements.

Extend »»

6 Astatine is radioactive. Suggest why it is difficult to study astatine.

Enquiry »»»

7 Use a reliable website or data book to find out the melting points of the halogens. Plot a bar chart of the data. You will need to decide on a sensible scale for the *y*-axis

≫ Extend: Making predictions

Patterns in periods

It is much harder to spot patterns in the periods of elements. This is because elements with similar properties are grouped together in the vertical groups, so as you go across a period, the elements have very little in common in terms of their properties.

Li	Be
Lithium	Beryllium

B	C	N	O	F	Ne
Boron	Carbon	Nitrogen	Oxygen	Fluorine	Neon

However, you can still find patterns in the chemical and physical properties of the elements in a period, especially if you know about those properties in the period above or below. Here are some data for the elements in Period 2.

Element	Li	Be	B	C	N	O	F	Ne
Metal or non-metal?	M	M	NM	NM	NM	NM	NM	NM
Melting point (°C)	181	1287	2076	Approx. 5530	−201	−219	−220	−249
Boiling point (°C)	1330	2469	3927	Approx. 5530	−196	−183	−188	−246

Tasks

1. For each element as you go across Period 2, deduce whether it is a solid, liquid or gas at room temperature (25 °C).
2. Compare the melting point and boiling point for carbon. What will happen to carbon as it is heated constantly from room temperature to a temperature of 6000 °C? Explain your answer.
3. Plot a graph of the melting points of the elements in Period 2. You will need to choose which type of graph to draw.
4. Plot a graph of the boiling points of the elements in Period 2.

The elements in Period 3 are listed in the table below.

Element	Na	Mg	Al	Si	P	S	Cl	Ar
Metal or non-metal?								
Solid, liquid or gas?								

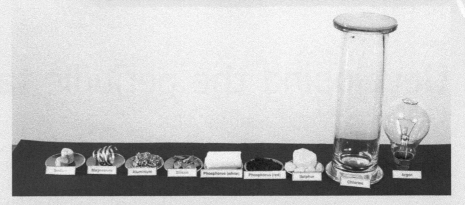

▲ The elements in Period 3. Two forms of phosphorus are shown

Tasks

⑤ Copy the table and make predictions to allow you to fill in the missing cells.
⑥ Which element in Period 3 do you think will have the highest melting point? Explain your answer.
⑦ Which element in Period 3 do you think will have the lowest boiling point? Explain your answer.

Predictions within a group

Look at the boiling points of the noble gases. We can use the trend in the data to suggest a value for the boiling point for argon. You would never be expected to estimate the exact correct answer, but you should still give a specific negative number for your prediction.

Element	Boiling point (°C)
Helium	−269
Neon	−46
Argon	
Krypton	−153
Xenon	−108

Tasks

⑧ Estimate a value for the boiling point of argon.
⑨ Do you think that argon will be chemically reactive or not? Explain why.
⑩ A scientist was given a sample of an element and asked to identify it. The element was a metal. It was soft and reacted quickly with the air. When she added it to water, it floated and reacted quickly, releasing hydrogen and producing an alkaline solution. What conclusions could the scientist make about the element?
⑪ What further tests would she need to do to identify exactly which element it was?

Enquiry:
Developing the periodic table

We take the periodic table for granted in science lessons these days, and so did your parents and grandparents when they were at school. But it took scientists many years to develop the periodic table, and in order to build it they had to make sense of patterns in the physical properties and chemical properties of the elements that they knew about at the time.

Changing ideas over time

The idea of what an element actually is had changed a lot over the time since the Ancient Greeks. But even when scientists decided to define an element as a substance that cannot be broken down into simpler substances by a chemical reaction, there was still some confusion. An early list of elements from a French chemist called Lavoisier included oxygen, nitrogen, hydrogen, phosphorus, mercury, zinc, sulfur, light and calorific (heat). We now know that these last two are types of energy rather than elements!

Döbereiner's triads

▲ Döbereiner, 1780 –1849

In 1817, German chemist Döbereiner noticed that amongst the elements that were known at the time, there were groups of elements which had similar chemical and physical properties. These groups seemed to contain **three** elements, so he called these groups **triads**. One of his triads was chlorine, bromine and iodine. Another was lithium, sodium and potassium.

Today, we recognise that these groups of elements are all in the same **group** in the modern periodic table. However, Döbereiner's model wouldn't be very helpful when other elements were later discovered which also belonged in the same group.

Newlands' Law of Octaves

In 1864, the British chemist John Newlands arranged the elements in order of increasing atomic weight. Atomic weight was an estimate of how heavy the atoms of that element were. He noticed a repeating pattern in the chemical and physical properties of every eighth element, so he called this his Law of Octaves, just like the musical scale of notes. Other scientists laughed at Newlands, and his work was not recognised by the Royal Society of Chemistry.

▲ Newlands, 1837–1898

Mendeleev's Periodic Table

▲ Mendeleev, 1834–1907

In 1869, Russian chemist Mendeleev published his periodic table. It incorporated the grouping of similar elements like Döbereiner's triads, and also the repeating periodicity of Newlands' Law of Octaves. But the most important thing about Mendeleev's periodic table was that he left gaps where he thought there was an element which hadn't been discovered at the time. He even used his periodic table to make predictions about the properties of these unknown elements, and he turned out to be right about most of them! He didn't make any predictions about the noble gases though, and these had to be added as a new group later on.

			K = 39	Rb = 85	Cs = 133	—	—
			Ca = 40	Sr = 87	Ba = 137	—	—
			—	?Yt = 88?	?Di = 138?	Er = 178?	—
			Ti = 48?	Zr = 90	Ce = 140?	?La = 180?	Th = 231
			V = 51	Nb = 94	—	Ta = 182	—
			Cr = 52	Mo = 96	—	W = 184	U = 240
			Mn = 55	—	—	—	—
			Fe = 56	Ru = 104	—	Os = 195?	—
			Co = 59	Rh = 104	—	Ir = 197	—
Typische Elemente			Ni = 59	Pd = 106	—	Pt = 198?	—
H = 1	Li = 7	Na = 23	Cu = 63	Ag = 108	—	Au = 199?	—
	Be = 9,4	Mg = 24	Zn = 65	Cd = 112	—	Hg = 200	—
	B = 11	Al = 27,3	—	In = 113	—	Tl = 204	—
	C = 12	Si = 28	—	Sn = 118	—	Pb = 207	—
	N = 14	P = 31	As = 75	Sb = 122	—	Bi = 208	—
	O = 16	S = 32	Se = 78	Te = 125?	—	—	—
	F = 19	Cl = 35,5	Br = 80	J = 127	—	—	—

▲ Mendeleev's early periodic table (1869). He later changed the layout to that used today, with periods arranged in horizontal rows

1. Why did Döbereiner group lithium, sodium and potassium together in the same triad?
2. Suggest (or research) why Newlands' work was laughed at by other scientists at the time, despite being a useful model.
3. Suggest why the noble gases were not discovered until much later than most of the other elements.
4. Imagine that you are a friend and fellow scientist of Newlands. His work has just been rejected by the Royal Society of Chemistry and he is feeling sad. Write a letter to Newlands to cheer him up. Include scientific detail about why his ideas are good, referring to the properties of specific elements.

Your knowledge objectives

In this chapter you will learn:
- that most substances are not pure elements, but compounds or mixtures, which have different properties to the elements they contain
- how to name simple compounds
- the symbols of some common elements
- the definitions of the terms element, atom, molecule, compound, chemical formula and polymer

See page 99 for the full learning objectives.

» Transition: What is an element?

Many thousands of years ago, Ancient Greek philosophers thought that everything was made from four elements: earth, air, fire and water. A few hundred years ago, scientists began to understand that these were not elements at all.

Elements

Elements are pure substances which are listed on the periodic table. Elements are made from just one type of atom. You will learn more about what atoms are when you begin your GCSE course.

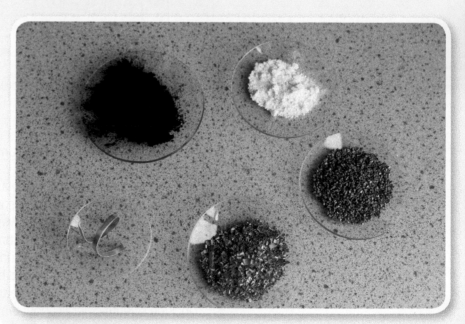

▲ All of these substances are elements. Clockwise from top left: carbon, sulfur, iron, copper, magnesium

Compounds

Compounds are pure substances that are made from two or more elements which are chemically combined in a fixed ratio of atoms. Because the ratio of atoms is fixed, a compound has a **chemical formula**. Water has the formula H_2O. This means that each water molecule is made from two hydrogen atoms and one oxygen atom.

Key words

A **compound** is a pure substance made from two or more elements which are chemically combined in a fixed ratio of atoms.

The **chemical formula** of a substance tells you which elements are present, and how many atoms of each there are.

▲ All of these substances are compounds. Clockwise from top left: iron chloride, copper sulfate, potassium iodide, cobalt nitrate, potassium manganate, sodium chloride

Properties of elements and compounds

The properties of a compound are different from the properties of the elements that it is made from. For example, water (H_2O) is made from the two elements hydrogen (H_2) and oxygen (O_2). Hydrogen and oxygen are both gases, but water is a liquid.

Worked example

Sodium chloride is a compound which is made from sodium and chlorine. Sodium is a soft metal. Chlorine is a green gas. Is it possible to use this information to predict the appearance of sodium chloride? Explain your answer.

No, you cannot predict the appearance of sodium chloride from this information. This is because sodium chloride is a compound made from the elements sodium and chlorine. The properties of a compound are always different from the properties of the elements it is made from.

Apply ≫

1 Why is water not listed on the periodic table?

2 Steel is an alloy made from iron, carbon and some other elements. It does not have a fixed chemical formula. Is steel an element, a compound or a mixture?

3 How many of the Ancient Greek 'elements' are actually elements? How many are compounds?

» Core: Representing elements and compounds

Elements and compounds

Elements are pure substances which are listed on the periodic table. An element is made from billions of tiny identical particles called atoms. We often represent atoms as small circles or spheres in diagrams.

Solid element   Liquid element   Gas element

▲ Atoms of an element can be arranged in a solid, liquid or gas

Compounds are pure substances which are made from two or more elements chemically bonded together. It is hard to separate the elements in a compound, and you will need a chemical reaction to do this. It is much easier to separate the substances in a mixture, as we saw in Pupil's Book 1, Chapter 10.

Atoms and molecules

The smallest part of a living organism is a cell, and biologists describe cells as the building blocks of living organisms. For chemists, the building blocks of substances are atoms. An atom is the smallest particle of an element that can exist. It is much, much smaller than anything else you have studied in science. A single strand of human hair is about 1 million atoms wide. Only the world's most powerful microscopes can 'see' individual atoms.

Sometimes, atoms bond to each other in clusters. A cluster of atoms like this is called a **molecule**. If the atoms in a molecule are identical, it is a molecule of an element. If the atoms in a molecule are different, it is a molecule of a compound. Some molecules are very small, but others may contain thousands of atoms.

Molecules of an element   Molecules of an element   Molecules of a compound   Molecules of a compound

> **Key word**
>
> A **molecule** is a tiny particle of a compound. Molecules are made from atoms which are strongly bonded together.

> **Key word**
>
> A **polymer** is a very long molecule made from thousands of smaller molecules joined together in a repeating pattern.

Polymers

A **polymer** is a very long molecule which typically contains thousands of atoms bonded together. Polymers are made from smaller molecules which are joined to each other in a repeating pattern, like the links in a chain or the beads on a very long necklace. All plastics are man-made polymers. Naturally occurring polymers include starch, DNA and proteins.

Giant structures

Some substances are not made from molecules at all. Sometimes, atoms bond to each other in giant structures. Examples include diamond, salt (sodium chloride) and all metals.

▲ The black balls represent the carbon atoms in diamond, which has a giant structure. The grey sticks represent strong chemical bonds

Worked example

What can you deduce about the substance indicated in the diagram? Use key words in your description.

The diagram shows a compound because it has more than one type of atom/element. The atoms are arranged in molecules. It is a gas.

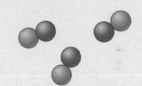

Know >

1 What is the smallest particle of an element called?

2 What name is given to a pure substance made from three elements?

3 Name two naturally occurring polymers and one man-made polymer.

Apply »

4 Which is larger, a red blood cell or a carbon atom?

5 What can you deduce about the substance shown in the diagram on the right?

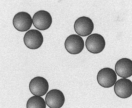

Extend »»

6 What can you deduce about the substance shown in the diagram below?

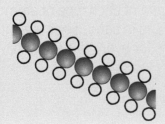

7 What can you deduce about the substance shown in the diagram to the right? Identify the state of matter in your answer.

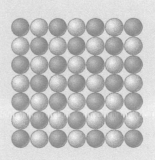

» Core: Interpreting chemical formulae

The language of chemistry

You can always refer to the periodic table to find out the symbol for an element, but it is useful to know some of the most common ones. The table below shows some of the common metals on the left and non-metals on the right.

Cu	copper	H	hydrogen
Zn	zinc	O	oxygen
Fe	iron	N	nitrogen
Al	aluminium	C	carbon
Na	sodium	S	sulfur
K	potassium	Cl	chlorine
Mg	magnesium	Br	bromine
		I	iodine

Formulae of molecules

Water

Methane

A **chemical formula** tells us which elements are present in a compound and the ratio in which the atoms of those elements are present. The simplest examples are for compounds that are made from molecules. Water has the formula H_2O, which tells you that there are two hydrogen atoms and one oxygen atom in each molecule. Methane has the formula CH_4, so it has one carbon atom and four hydrogen atoms in each molecule.

Elements can have a formula too, if they are made from molecules. Oxygen has the formula O_2, which means that each molecule is made from two oxygen atoms. Sulfur atoms are usually bonded in molecules of eight, so the formula is S_8.

Formulae of giant structures

Not all compounds are made from molecules. Some compounds have a giant structure instead. For giant structures, we use the chemical formula to show the simplest ratio of each type of atom present. Silicon oxide is the main compound in sand. It has a giant structure but there are two oxygen atoms for every silicon atom, so the formula of silicon oxide is SiO_2.

▲ In this model of silicon oxide, there are twice as many oxygen atoms (red) as silicon atoms (black) so the formula is SiO_2

Naming compounds

When elements combine to make a compound, the name of the compound can usually be worked out from the names of the elements. If a metal is present, that comes first. Then the name of the non-metal element usually changes to end in -ide.

For example:

sodium + chlorine → sodium chloride

aluminium + iodine → aluminium iodide

oxygen + magnesium → magnesium oxide.

The prefixes mono-, di- and tri- are also used to show that a compound contains one, two or three of a specific type of atom. For example:

- carbon monoxide (CO) has one oxygen atom

- carbon dioxide (CO_2) has two oxygen atoms

- sulfur trioxide (SO_3) has three oxygen atoms.

You should also know that some compounds contain groups of atoms which are often found together in a compound.

Group of atoms	Compound name will include...	Example
OH	Hydroxide	Sodium hydroxide, NaOH
NO_3	Nitrate	Potassium nitrate, KNO_3
SO_4	Sulfate	Copper sulfate, $CuSO_4$
CO_3	Carbonate	Calcium carbonate, $CaCO_3$

Worked example

State the names of the elements present in the compound with the formula K_2SO_4. How many of each type of atom are present?

The compound contains potassium, sulfur and oxygen. There are two potassium atoms, one sulfur atom and four oxygen atoms.

Know >

1 What is the symbol for sulfur?

2 Which element has the symbol Na?

3 Which group of atoms will always be present in a compound that has 'nitrate' in its name?

Apply >>

4 State the number of each type of atom present in aluminium oxide (Al_2O_3).

5 What is the name of the compound with the formula NaBr?

6 What is the name of the compound with the formula FeS?

7 What is the name of the compound with the formula $FeSO_4$?

Extend >>>

8 Lithium, sodium and potassium are all elements in Group 1 and react to form similar compounds. The formula of sodium hydroxide is NaOH. The formula of potassium hydroxide is KOH. Suggest the formula of lithium hydroxide.

9 The formula of sodium carbonate is Na_2CO_3. Suggest the formula for lithium carbonate and potassium carbonate.

10 Some compounds have a name which doesn't tell you anything useful about the elements it is made from, or the ratio of the atoms. Water has the formula H_2O. If you had to name water 'properly' using the rules described on these pages, what would you call it?

▲ The plate, mug and teapot are made from a ceramic material

» Extend: Ceramics and composites

Ceramics and composites are substances which have very useful properties, but they are quite different from simple elements and simple compounds. There are many examples of ceramics and composites all around us.

Ceramics

The most familiar ceramic items in our daily lives are the china plates and mugs that we use to eat and drink from. The word 'ceramic' comes from the Greek word for pottery, and most ceramics are man-made, using strong heating as part of the manufacturing process.

Ceramics usually have a very high melting point, are brittle, and poor conductors of heat and electricity compared with other solids such as metals. Silicon carbide ceramics can be used to make brake discs in supercars. The friction caused by the brake pads squeezing the brake disc when the driver brakes hard to slow down from a high speed makes these ceramic brake discs get very hot. Red hot in fact! But the brake discs do not melt or change shape because they are made from this special ceramic.

▲ The brake discs are made from a ceramic so they do not melt, even when they get red hot

Composite materials

A composite is a material which is made from two other materials which have different physical properties. The composite will typically have the best of both materials. Many composites are man-made, but bone is one example of a naturally occurring composite. Here are some examples of composites.

Composite	Made from		Useful properties	
Reinforced concrete	Concrete	Strong but brittle	Strong and flexible	
	Steel	Flexible, strong, heavy		
Bone	Calcium phosphate	Strong but brittle	Strong and flexible	
	Protein	Flexible		
Carbon fibre reinforced polymer	Carbon fibre	Strong, light and flexible	Very strong, stiff and lightweight	
	Epoxy resin	Brittle, heavy and stiff		

Tasks

1. Suggest why ceramic tiles are used to coat the surface of a spacecraft to enable it to protect the astronauts as it re-enters the Earth's atmosphere.
2. Which composite listed in the table would be used for the frame of a racing bicycle? Explain your answer.
3. Plywood is a composite made from many thin layers of wood, glued together so that the wood grain is in different directions. Suggest why plywood is a better material for building furniture than normal wood.

Enquiry:
Element or compound?

Chemists define an element as a pure substance that is made from one type of atom. However, the kind of powerful microscope that has allowed us to 'see' individual atoms (called a scanning tunnelling microscope) was only invented in 1981 and scientists knew all about atoms long before that!

▲ An image of some palladium atoms (shown in white) on top of a sheet of carbon atoms (shown in blue) made using a scanning tunnelling microscope. The colours have been added afterwards by computer

An important part of the definition of an element was that it was a pure substance that could not be broken down into simpler substances using chemical reactions. This helped early chemists investigate substances and work out whether they were elements or not. Here are two experiments you can do to a substance to see if it is an element. You might be able to try them in the school laboratory.

Heat it up to a very high temperature in air

Does it break down into simpler substances? It is definitely not an element! Calcium carbonate is a compound which breaks down into two simpler substances when it is heated.

Does it burn? It might be an element, like carbon or sulfur, which both form gases when they burn. Many compounds burn too. Metals often burn, and they form solid metal oxides, which are heavier.

▲ Sulfur burns in air with a blue flame to form sulfur dioxide gas

Dissolve it in water or melt it, and then pass an electric current through the liquid

This process is called **electrolysis** and was invented by Martinus van Marum in 1785. Humphry Davy used this process to discover lots of metal elements in 1808. Electrolysis produces elements from compounds. For example, you can dissolve the compound copper chloride into water, electrolyse it and then carefully smell the tiny bubbles of chlorine gas produced at the positive electrode (you will recognise the smell of swimming pools or bleach). The negative electrode turns a red-brown colour because of the copper metal formed there.

Key word

Electrolysis is when a compound is broken down into elements using electricity.

▲ The electrolysis of copper chloride solution. Bubbles of chlorine can be seen on the right electrode during the experiment (a). After the experiment, copper can be seen on the left electrode (b)

1. In 1772, Swedish chemist Carl Scheele heated solid mercury oxide and discovered that it formed a gas and a liquid which was silver coloured and very heavy. Which two elements had he formed from the compound? Write a word equation for this reaction.
2. Zinc chloride is a white solid. It can be melted by heating strongly with a Bunsen burner. If two electrodes are put into molten (liquid) zinc chloride, two substances are made. One is a silvery metal and the other is a gas. Suggest what is happening in this experiment.
3. Aluminium powder is pale grey. Iodine is a solid made from grey crystals. When aluminium and iodine are mixed together and a drop of water is added, there is a white flame, purple smoke and a new substance is made, which is a white powder. Suggest the name of the new chemical product and state whether it is an element or compound.

Reactions

Learning objectives

11 Chemical energy

In this chapter you will learn...

Knowledge

- that during a chemical reaction bonds are broken (requiring energy) and new bonds are formed (releasing energy)
- that if the energy released is greater than the energy required, the reaction is exothermic; if the reverse, it is endothermic
- the definitions of the terms catalyst, exothermic, endothermic and chemical bond

Application

- how to use experimental observations to distinguish between exothermic and endothermic reactions
- how to use a diagram of relative energy levels of particles to explain energy changes observed during a change of state

Extension

- how to predict whether a chemical reaction will be exothermic or endothermic given data on bond strengths

12 Types of reaction

In this chapter you will learn...

Knowledge

- that combustion is a reaction with oxygen in which energy is transferred to the surroundings as heat and light
- that thermal decomposition is a reaction in which a single reactant is broken down into simpler products by heating
- how to describe what happens to atoms and chemical bonds during chemical reactions
- that during a reaction the total number of atoms is conserved
- the definitions of the terms fuel, chemical reaction, physical change, reactants, products and conserved

Application

- how to explain why a given reaction is an example of combustion or thermal decomposition
- how to predict the products of the combustion or thermal decomposition of a given reactant
- how to write word equations from information about chemical reactions
- how to explain observations about mass in a chemical or physical change
- how to use particle diagrams to show what happens in a reaction

Extension

- how to use known masses of reactants or products to calculate unknown masses of the remaining reactant or product
- how to balance a symbol equation
- how to use mass of reactant in an equation to determine mass of product, e.g. magnesium and oxygen

11 Chemical energy

» Transition: Energy in chemical reactions

Your knowledge objectives

In this chapter you will learn:
- that during a chemical reaction bonds are broken (requiring energy) and new bonds are formed (releasing energy)
- that if the energy released is greater than the energy required, the reaction is exothermic; if the reverse, it is endothermic
- the definitions of the terms catalyst, exothermic, endothermic and chemical bond

See page 123 for the full learning objectives.

When you light a Bunsen burner, you start off a chemical reaction. Energy is released when the methane from the gas supply burns. Because energy is released in this reaction, we describe it as an exothermic reaction.

Physical changes and chemical reactions

Some changes are easy to reverse, and don't involve new chemicals being made, or chemical bonds being broken or made. These changes are called physical changes. Physical changes include melting, freezing, boiling, evaporating and condensing. They also include dissolving and crystallisation.

> **Key word**
>
> In an **exothermic reaction**, energy is given out, usually as heat or light.

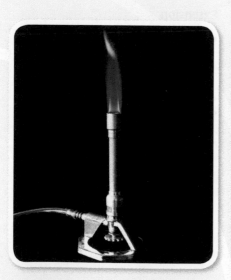

▲ When methane burns in a Bunsen burner flame, energy is released

▲ Melting and boiling are easy to reverse and don't make a new compound, so they are physical changes

Key word

In an **endothermic reaction**, energy is taken in, usually as heat.

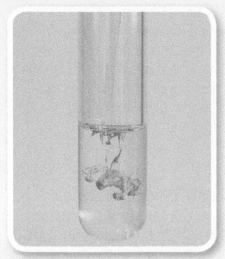

▲ In this chemical reaction, there is an obvious colour change and the reaction cannot be reversed

Chemical reactions are usually much harder (or impossible) to reverse. They involve chemical bonds between atoms being broken and made. New compounds are made in chemical reactions.

Some reactions need to take in lots of energy to make them happen. We call this type of reaction an **endothermic reaction**.

Energy changes can also be described during physical changes, like changes of state. When ice is heated, it melts to form water. You can see this in the diagram below.

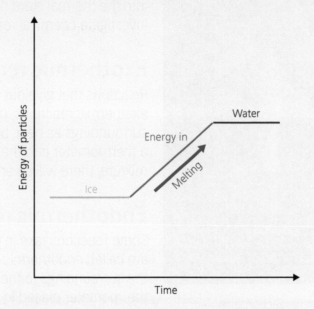

During melting and boiling, energy is absorbed. During condensing and freezing, energy is released from the chemicals into the surroundings.

Worked example

When a piece of wood burns in a fire, is this a chemical reaction or a physical change? Explain your answer.

This is a chemical reaction, because once the wood is burned, you cannot get it back. The reaction is irreversible, and new chemicals are made, like carbon dioxide and water.

Apply »

For each of the following questions, say whether it is a chemical reaction or a physical change and explain your answer.

1 Describe what happens when a student mixes some sand into some water and stirs.

2 Describe how to turn liquid bromine into bromine gas.

3 How can solid sodium chloride be obtained from sea water, which is a solution of sodium chloride?

» Core: Endothermic and exothermic reactions

Energy changes in reactions

Most of the chemical reactions that you have seen in science lessons have released energy. Sometimes this energy is easy to notice, because if you hold the boiling tube then your hand feels warm, as energy moves from the chemical store of the chemicals into the thermal store of your hand. Sometimes, you get to investigate chemical reactions that feel cold.

Exothermic reactions

Reactions that give out energy to the surroundings are called exothermic reactions. The energy is usually transferred to the surroundings as heat, but sometimes as light or sound. If you put a thermometer into the reaction mixture, or near to the reaction mixture, there will be an increase in temperature.

Endothermic reactions

Some reactions take in energy from the surroundings and they are called endothermic reactions. If this energy is transferred from the surroundings to the chemicals as heat, then the reading on a thermometer placed in the reaction mixture or the surroundings will show a decrease in temperature.

▲ The thermite reaction is an exothermic reaction which releases a lot of energy

Common error

Some people think that if energy is going from the surroundings into the chemicals then they should get hot. But when the energy is absorbed from the surroundings, it goes into the chemical energy store of the chemicals, not their thermal store.

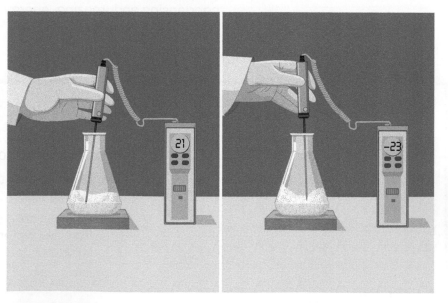

▲ In this endothermic reaction, the temperature at the start is 21 °C and at the end it is –23 °C!

Interpreting results from experiments

A student investigated four experiments to see whether they were exothermic or endothermic. Here are her results.

	Experiment	Temperature at start (°C)	Temperature at end (°C)
1	Add magnesium ribbon to hydrochloric acid	21	45
2	Add water to solid ammonium nitrate	23	5
3	Add sodium carbonate to ethanoic acid	22	11
4	Add solid copper sulfate to water	23	33

From the data in the table, we can see that Experiment 1 was exothermic, because there was a rise in temperature. The temperature change is calculated by subtracting the start temperature from the end temperature. So, 45 − 21 = 24 °C.

We can also see that Experiment 2 was endothermic, because there was a temperature change of −18 °C.

Worked example

A student put 5 cm³ water into a boiling tube and recorded the temperature. It was 18 °C. He then added one spatula of calcium metal granules and found that after 30 s, the temperature of the mixture was 39 °C. Explain whether the reaction was exothermic or endothermic.

The reaction was exothermic because thermal energy was released into the surroundings, causing the temperature of the mixture to rise.

Know >

1 What is the definition of an exothermic reaction?

2 What is the definition of an endothermic reaction?

3 Name three types of energy that can be transferred in a chemical reaction.

Apply >>

4 On Bonfire Night, glow sticks are very popular. When they are activated, a reaction makes the chemicals inside the glow stick glow with a coloured light. Is this an exothermic or endothermic reaction? Explain your answer.

Extend >>>

5 Is photosynthesis exothermic or endothermic? Explain your answer.

6 Look at the results in the table above. Two of the experiments were actually physical changes and not chemical reactions at all. Which two? (Note that the terms endothermic and exothermic can be used to describe physical changes as well as chemical reactions.)

Enquiry >>>>

7 For Experiments 3 and 4 in the table, calculate the temperature change and deduce whether they were exothermic or endothermic.

» Core: Explaining energy changes

Breaking and making bonds

Key word

Chemical bonds are strong forces that hold two atoms together.

During any chemical reaction, the first stage involves breaking the **chemical bonds** in the reactant molecules. Once this has been done, new bonds can form between the atoms, in order to make the product molecules. We can model this with a simple reaction, like hydrogen and oxygen reacting to form water. Here is the overall reaction:

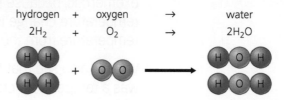

Here is the first step – breaking all the bonds in the reactant molecules. This step requires energy to be absorbed from the surroundings, so energy goes in. This is easy to remember: 'breaking in', like a new pair of shoes.

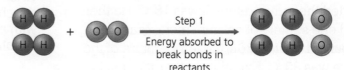

Then the reaction finishes when new bonds are made between the atoms in the product molecules. This step releases energy, so energy comes out. This is easy to remember: 'making out', like in a teenage relationship. So, we can say that explaining energy changes can be done with the mnemonic 'Shoes and snogging: breaking in, and making out'.

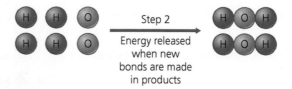

What determines whether a reaction is going to be exothermic or endothermic is the amount of energy absorbed in step 1 and the amount of energy released in step 2. If a little bit of energy is absorbed to break bonds in the reactants, and then a lot of energy is released when new bonds are made in the products, the reaction is exothermic overall. For an endothermic reaction, a lot of energy is absorbed to break reactant bonds, and then only a little bit of energy is released when bonds are made in the products. We can show this on an energy level diagram.

Exothermic

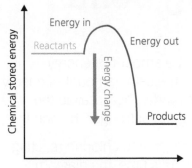

Endothermic

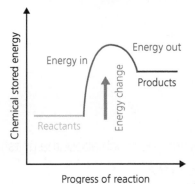

Energy level diagrams

During the reaction, we move from left to right along the diagram. An exothermic reaction has less chemical energy stored in the products than it had stored in the reactants, which explains where the energy was released from. There is a rise in energy to begin with, as the chemicals absorb energy initially to break bonds in the reactants, and then they release energy later on in the reaction. In an endothermic reaction, the products have more chemical energy than the reactants, but there is still a rise in energy and then a fall in energy.

Catalysts

A **catalyst** is a chemical which speeds up a reaction but is not used up or chemically changed at the end of the reaction. Catalysts are very important in the chemical industry because making chemicals more quickly means you can sell more of them for profit. It also means you can use a lower temperature, which takes less energy and saves resources. Catalysts are found inside all our cells too. When a catalyst is inside a living cell it is called an **enzyme**.

Catalysts work by reducing the amount of energy needed to break bonds in the reactants. This means that the reaction can get going more quickly, without so much energy having to be absorbed. We can show this on an energy level diagram.

The energy change of the reaction is unchanged, but it can now go faster because less energy is needed to break the reactant bonds.

Key words

A **catalyst** is a chemical which speeds up a reaction but is not used up or chemically changed.

Enzymes are proteins made by living organisms to speed up chemical reactions that take place within cells.

Worked example

When methane reacts with oxygen in a Bunsen burner to produce carbon dioxide and water, the reaction is exothermic. Explain why, using ideas about breaking and making bonds.

Energy is absorbed to break bonds between the atoms in the methane and oxygen molecules. When the new bonds are made in the carbon dioxide and water, energy is released. More energy is released when making the new bonds in the products than was absorbed to break the bonds in the reactant molecules.

Know >

1 What is the definition of a catalyst?

2 In a chemical reaction, what happens first: breaking or making bonds?

Apply >>

3 When potassium carbonate reacts with nitric acid to form potassium nitrate, water and carbon dioxide, the reaction is endothermic. Explain why, using ideas about breaking and making bonds.

Extend >>>

4 On the diagrams above, the energy needed to break the bonds in the reactants is shown using a blue line. Using this information, and the information on catalysts above, draw an energy level diagram to show the effect of a catalyst on an exothermic reaction.

» Extend: Using bond energies

We can find out from a reliable source the amount of energy needed to break a specific bond. This is the same amount of energy that will be released when that bond is made. This is called the bond energy. The higher the bond energy, the stronger the bond.

Let's look at a simple reaction, like hydrogen and chlorine reacting together to form hydrogen chloride gas. This time, the diagram shows the chemical bonds between the atoms, which are all represented with lines. Here is the overall reaction.

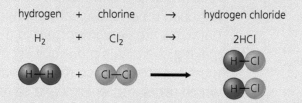

| hydrogen | + | chlorine | → | hydrogen chloride |
| H_2 | + | Cl_2 | → | 2HCl |

Here is a table of bond energies. (Don't worry too much about the units!)

Bond	Bond energy (kJ/mol)
H–H	436
Cl–Cl	242
H–Cl	432

The first step in the reaction is to break the H–H bond and also the Cl–Cl bond. The total amount of energy needed to do this is 436 + 242 = 678 kJ/mol. The atoms now look like this:

The final step in the reaction is for the new H–Cl bonds to form in the product molecules. Two of these bonds will form, to make the two separate HCl molecules you can see in the top diagram. So, the amount of energy released when two H–Cl bonds form is 2 × 432 = 864 kJ/mol.

Here are the product molecules:

The total energy change for this reaction can be calculated by remembering that the energy in at the start of the reaction is positive on our energy level diagram, but the energy released when the new bonds are made is negative (down) on our energy level diagram. So, to calculate the energy change, we have to subtract the energy released from the energy absorbed.

energy change = energy absorbed – energy released

energy change = 678 – 864 = –186 kJ/mol

The fact that this value is negative tells us that the reaction is exothermic. An endothermic reaction would have a positive energy change. Look back at the section on energy level diagrams to see why.

A bond diagram for the reaction between hydrogen and oxygen to form water is shown below. You can see that the oxygen atoms are joined by a double bond.

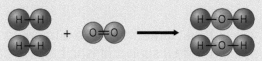

In this reaction, the first step is to break two H–H bonds and one O=O (we are breaking one double bond here).

Tasks

1. Identify the number and type of each bond being made in the products.
   Here is a table of bond energy values.

Bond	Bond energy (kJ/mol)
H–H	436
O=O	498
H–O	459

2. Calculate the energy that must be absorbed to break two H–H bonds and one O=O bond.
3. Calculate the energy which is released when the bonds in the products are made.
4. Calculate the energy change of the reaction, by subtracting your value for Question 3 from your value for Question 2.

Enquiry:
Handwarmers and cold packs

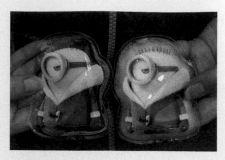

▲ A reusable handwarmer before being activated (left) and after (right)

Handwarmers

Have you ever used a handwarmer? These use a chemical reaction to release energy and this is what keeps your hands warm. Some are reusable, because after they have been used they can be recharged, by putting them in boiling water for 10 minutes and then allowing them to cool down. Other types can only be used once.

1 Why would a reusable handwarmer be better for the environment?

Ravi wanted to make a handwarmer by mixing two chemicals inside a sealed plastic bag. He decided to test five reactions to see which one would make the best handwarmer. Here are his results.

	Reaction	Start temp. (°C)	End temp. (°C)	Temp. change (°C)	Observations
1	Calcium oxide + water	21	95		Calcium hydroxide is produced, which is corrosive.
2	Copper sulfate + water	22	49		Copper sulfate is harmful.
3	Sodium ethanoate crystallising	20	41		This process can be reversed.
4	Calcium + nitric acid	18	55		Bubbles of gas were produced.
5	Sodium carbonate + ethanoic acid	19	14		Bubbles of gas are produced. All reactants and products are low hazard.

2 Calculate the temperature change for each reaction.
3 Which one of the reactions is an endothermic process?
4 Other than the fact that the product (calcium hydroxide) is corrosive, suggest why it would be dangerous to use Reaction 1 in a handwarmer for children.
5 Reaction 4 produced bubbles of a gas. Identify this gas by looking at the reactants and explain why this might be a dangerous reaction.
6 Ravi decided that Reaction 3 would be the best for his handwarmer. Give two reasons why.

Cold packs

Instant cold packs can be used to treat injuries such as strains and sprains. They help to ease pain and reduce swelling.

Inside the strong plastic bag are lots of crystals of ammonium nitrate. There is also a small plastic bag full of water. The small plastic bag is not very strong, so you can burst it by squeezing it with your fingers. This allows the water and ammonium nitrate to mix inside the strong outer bag, but you don't touch the chemicals.

Archie and Ravi decided to investigate the perfect amount of water and ammonium nitrate to mix in an instant cold pack which would reach the lowest temperature.

▲ Instant cold packs can be used to treat injuries

▲ The contents of an instant cold pack

?

7 What kind of reaction is needed in an instant cold pack, exothermic or endothermic?
8 Use ideas about energy to explain why.

Archie suggested that they try changing the mass of ammonium nitrate and the volume of water, and measure the temperature of the solution after 10 minutes. Ravi said that it would be better if they kept the mass of ammonium nitrate the same each time, and that they should record the lowest temperature the mixture reached, not the temperature after 10 minutes.

?

9 Explain why Ravi's ideas would make this a better investigation.
10 What is the dependent variable in this investigation?

Archie and Ravi measured 10g of ammonium nitrate into five boiling tubes. They added a different volume of water to each tube and recorded the lowest temperature each tube got to with some gentle stirring. Here are their results.

Volume of water added	Lowest temperature
3	12
6	5
9	10
12	13
15	16

?

11 What is missing from their results table?
12 Plot a graph of their results. Draw a smooth curve of best fit to show the trend in the data. Your curve does not have to go through all your points but it might do.
13 Describe the trend in the results.
14 What volume of water should they use in a cold pack that contains 100g of ammonium nitrate to make the most effective cold pack?

Types of reaction

Your knowledge objectives

In this chapter you will learn:
- that combustion is a reaction with oxygen in which energy is transferred to the surroundings as heat and light
- that thermal decomposition is a reaction in which a single reactant is broken down into simpler products by heating
- how to describe what happens to atoms and chemical bonds during chemical reactions
- that during a reaction the total number of atoms is conserved
- the definitions of the terms fuel, chemical reaction, physical change, reactants, products and conserved

See page 123 for the full learning objectives.

Key words

Combustion is another name for burning. This is when a fuel reacts with oxygen and releases heat and light energy.

Mass is **conserved** in a chemical reaction because it stays the same. The total mass of atoms before and after the reaction is the same.

Thermal decomposition is when a single reactant breaks down into simpler substances when it is heated.

» Transition: Burning and breaking down

Have you learned about the fire triangle before? Firefighters use the fire triangle to help teach people about how to prevent or deal with fires. The fire triangle identifies the three things that are needed for combustion (burning).

So you can put out a fire by preventing oxygen reaching it. This is how carbon dioxide fire extinguishers work. You can also put out a fire by cooling it down. Water from a sprinkler system or a firefighter's hose does this. And if a fire runs out of fuel, it will go out.

When things burn, they often seem to get lighter. At first, you might think that some of the atoms are being destroyed, but this is impossible. All of the atoms at the start of the reaction are still there at the end – we say that they are **conserved**. The reason that the ash left over is often lighter than the fuel was, is that lots of the atoms from the fuel are now in the air as carbon dioxide and water vapour.

▲ The fire triangle. If one of the sides is missing, combustion is not possible

▲ The ash weighs less than the wood did, but the mass has still been conserved. Atoms have moved from the wood into the air in the smoke, so the ash is lighter than the wood

Thermal decomposition

There are lots of chemicals that do not burn when they are heated. Some of them do not change at all. But some chemicals do change when they are heated. They break down into simpler substances. We call this kind of reaction **thermal decomposition**.

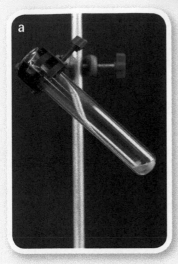
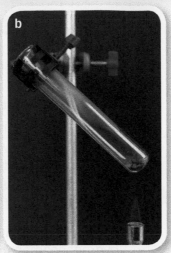
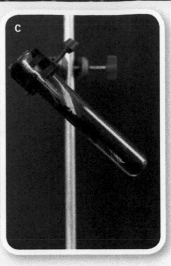

▲ Green copper carbonate (a) breaks down when it is heated (b) to form colourless carbon dioxide gas and black copper oxide (c)

Copper carbonate breaks down when it is heated. It makes copper oxide and carbon dioxide. Here is a word equation and symbol equation.

copper carbonate → copper oxide + carbon dioxide

$$CuCO_3 \rightarrow CuO + CO_2$$

You can see that there is only one reactant (on the left of the arrow). You can also see that all the atoms at the start are still there at the end.

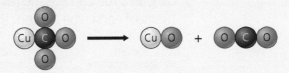

Worked example

Look at the following symbol equation for an exothermic reaction. Identify the type of reaction taking place and explain your answer.

$$CH_4 + 2O_2 \rightarrow CO_2 + 2H_2O$$

This reaction is a combustion reaction. I know this because a chemical is reacting with oxygen.

Apply »

1 Which element is always a reactant in a combustion reaction?

2 If someone's clothes catch on fire in a chemistry lesson, you can wrap them in a large fire blanket to put out the flames. It is a bit like a large towel made from a non-flammable material. Use the fire triangle to explain how it works.

3 Look at the following reaction, which takes place at a high temperature, and identify the type of reaction taking place. Explain your answer.

$$\text{calcium carbonate} \rightarrow \text{calcium oxide} + \text{carbon dioxide}$$

» Core: Thermal decomposition

Baking cakes

You will probably have made muffins at some point. If you're a good baker, you will have made muffins that were light and not too heavy. This is because when they were baking in the oven, bubbles of gas were produced inside your muffin mix which left behind air spaces in the cooked muffins.

One of the ingredients that goes into a muffin mix is sodium hydrogencarbonate. The common name for this is bicarbonate of soda, but chemists prefer the proper name because it tells us which elements are present:

sodium hydrogencarbonate

$NaHCO_3$

Sodium and hydrogen and carbon and oxygen.

The ending -ate tells us that oxygen is present in a compound.

When sodium hydrogencarbonate is heated, it breaks down. As we have already learned, this kind of chemical reaction is called thermal decomposition. Here is a word equation, symbol equation and particle diagram. You can see the names of the three products. Count how many of each type of atom there are at the start and after the reaction.

sodium hydrogencarbonate	→	sodium carbonate	+	carbon dioxide	+	water
$2NaHCO_3$	→	Na_2CO_3	+	CO_2	+	H_2O

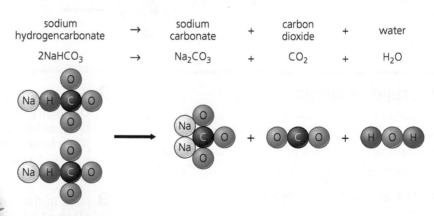

You can see that there are the same number of atoms present at the start and at the end. So we know that the total mass of the reactants is the same as the total mass of the products. This is always the case in **chemical reactions** (and **physical changes**). Mass is always conserved in a reaction. This means that it stays the same.

However, this doesn't seem like common sense in some situations. Sometimes, it seems as though the chemicals get lighter in a

▲ Eating muffins is more fun than learning about chemistry!

Key words

In **chemical reactions**, a new substance is made because chemical bonds are broken and made.

In **physical changes**, the properties of a substance change but no new substance is made. Changes of state and dissolving are all examples of physical changes.

reaction. If you weigh out 2g of sodium hydrogencarbonate and heat it strongly in a boiling tube for 5 minutes and then weigh it again, you will find that its mass has decreased. Have we broken the laws of science? Where have the atoms gone?!

Look at the products of the reaction in the equation. Carbon dioxide is a gas, and the water produced is a gas at this high temperature. So they escape from the boiling tube and when we weigh it at the end of the experiment, we are only weighing the glass boiling tube and the sodium carbonate – of course it gets lighter! If we could find a way to trap the two gases and weigh them too, then the numbers would all add up (to 2g).

State symbols can help us to make sense of mass changes in reactions:

- (s) means solid

- (l) means liquid

- (g) means gas

- (aq) means aqueous (dissolved in water)

When we use state symbols, the equation looks like this:

$$2NaHCO_3(s) \rightarrow Na_2CO_3(s) + CO_2(g) + H_2O(g)$$

> ### Key fact
> The law of **conservation of mass** states that the total mass of the reactants is the same as the total mass of the products.

Worked example

Copper carbonate powder decomposes at high temperatures to form copper oxide and carbon dioxide. Write a word equation to show this reaction.

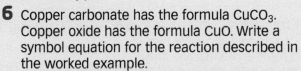

copper carbonate $\rightarrow$ copper oxide $+$ carbon dioxide

Know >

1 What is the definition of thermal decomposition?

2 Name the three elements present in sodium carbonate.

3 If you see Ca(s) in a chemical equation, what state is the calcium in?

Apply >>

4 Use ideas about atoms to explain why the total mass of the reactants must equal the total mass of the products.

5 Magnesium carbonate thermally decomposes when heated to form magnesium oxide and carbon dioxide. Write a word equation for this reaction.

Extend >>>

6 Copper carbonate has the formula $CuCO_3$. Copper oxide has the formula CuO. Write a symbol equation for the reaction described in the worked example.

7 Limestone is a rock which is quarried on a large scale in parts of the UK. The main compound in limestone is calcium carbonate, $CaCO_3$. Some of this calcium carbonate is heated strongly. Suggest the name and formulae of the two products from this reaction. Write a word and symbol equation for the reaction.

Enquiry >>>>

A student investigated the thermal decomposition of copper carbonate by heating up samples of it in a boiling tube. She started with the following masses of copper carbonate in separate boiling tubes: 1g, 2g, 3g, 4g, 5g. She measured the mass of the contents of the boiling tubes after heating them strongly for 5 minutes.

8 Identify the independent variable and dependent variable in this investigation.

9 Design a results table for this investigation.

Key word

A **fuel** is a substance that can be burned to release energy.

» Core: Combustion

Combustion is another word for burning. In a combustion reaction, a **fuel** reacts with oxygen. The oxygen usually comes from the air. Energy is released, so the reaction is exothermic. The energy which is released is mainly in the form of heat, but also light. We say that energy is transferred from the chemical store of the fuel and oxygen mixture, to the thermal store of the surroundings.

▲ Enjoying a combustion reaction in the woods!

Combustion reactions

The chemicals on the left-hand side of a chemical equation are called the **reactants**. This is because they react with each other. The chemicals on the right of the arrow are called the **products**. This is because they are produced by the reaction. For example:

Key words

Reactants are substances that react together in a reaction. They are always before the arrow in an equation.

Products are substances which are formed in a reaction. They are always after the arrow in an equation.

methane + oxygen → carbon dioxide + water

$$CH_4(g) + 2O_2(g) \rightarrow CO_2(g) + 2H_2O(g)$$

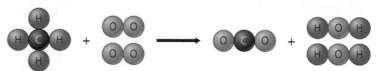

The two reactants are methane and oxygen. The two products are carbon dioxide and steam. All of the substances are gases in this reaction. In a combustion reaction, oxygen is always one of the reactants. The other reactant is the fuel.

Products of combustion

Lots of the fuels that we burn are made from carbon and hydrogen atoms. When they burn in a good supply of air, there are lots of oxygen atoms. This means that carbon dioxide and water are produced when many fuels are burned. If the air supply is bad, there are fewer oxygen atoms around. This means that carbon monoxide and water are produced, and less energy is released. Carbon monoxide is a poisonous gas, which is why it is important to make sure combustion reactions take place in a good air supply.

Mass changes in combustion

If the fuel being burned is a liquid or a solid, the mass seems to go down during a combustion reaction, because the products of most combustion reactions are gases. But because no atoms are created or destroyed, mass must always be conserved in a chemical reaction. That means that if you could trap the gaseous products and weigh them, the total mass of the products would be the same as the total mass of the reactants.

Worked example

When ethanol is burned in a camping stove, carbon dioxide and water are produced. Write a word equation for this reaction.

ethanol + oxygen → carbon + water
                              dioxide

Know >

1 Which gas is a reactant in every combustion reaction?

2 What name is given to a chemical which is burned to release energy?

3 Which two products are made during most combustion reactions, as long as there is a good supply of oxygen?

Apply >>

4 Write a word equation for the combustion of propane, to produce carbon dioxide and water.

5 Write a word equation for the combustion of butane in a good supply of oxygen.

6 Write a word equation for the combustion of butane in a poor supply of oxygen.

Extend >>>

7 Suggest why it would be very dangerous to take a lit barbeque into a tent at night, even if you were sure that it couldn't set fire to the tent.

8 Pure carbon can be burned as a fuel, when it reacts with oxygen. Work out what is produced by this reaction and write a word equation and symbol equation for it.

9 Hydrogen is described by many people as a clean fuel. Write a word equation for the combustion of hydrogen.

Enquiry >>>>

10 One of the products of combustion for most fuels is carbon dioxide. Describe how to test for this gas, and what the positive result is (i.e. what would you see happen if the gas you were testing really **was** carbon dioxide).

» Core: Balancing equations

Why do equations need to be 'balanced'?

We know that atoms cannot be made or destroyed. But when we change a word equation to a symbol equation and count up the number of each type of atom before and after the reaction, we often find that it doesn't make sense – it is not balanced. Have a look at this equation:

hydrogen + oxygen → water

$$H_2 + O_2 \rightarrow H_2O$$

Where has the second oxygen atom disappeared to?! We cannot change the formula of oxygen from O_2 to O. You must never change the little numbers in any chemical formula. We need to balance the equation. This is how to do it.

1. Draw a line under the equation using a ruler. Draw a vertical dotted line down from the arrow.	H_2 + O_2 → H_2O
2. Underneath the line, write out the atoms present in each reactant and product. This helps with counting the atoms.	H_2 + O_2 → H_2O HH OO HOH
3. Count the number of atoms of the first element. Check it is balanced on the left and right. In this case, the hydrogen is balanced, with two atoms on the left and the right.	H_2 + O_2 → H_2O HH OO HOH
4. Now count the number of atoms of the next element and check it is balanced. Here we can see that there are two oxygen atoms on the left and just one on the right.	H_2 + O_2 → H_2O HH OO HOH
5. Add another identical copy of the formula for water on the next line to balance the O atoms. You can only duplicate what is above. The oxygens are now balanced, but you will notice that the hydrogens are now no longer balanced.	H_2 + O_2 → H_2O HH OO HOH HOH
6. Adding another HH on the left balances the equation. Now we just need to count the number of each type of atom. Count the number of lines you have used in your book underneath the original equation.	H_2 + O_2 → H_2O HH OO HOH HH HOH
7. The large numbers in front of the formula show us how many of each molecule are needed.	$2H_2$ + O_2 → $2H_2O$ HH OO HOH HH HOH

You can see from the particle diagram that the equation is now balanced:

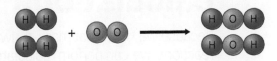

Apply »

Here are some equations for you to balance:

1 $H_2O_2 \rightarrow H_2O + O_2$

2 $CH_4 + O_2 \rightarrow CO_2 + H_2O$

3 $Mg + O_2 \rightarrow MgO$

4 $C_3H_8 + O_2 \rightarrow CO_2 + H_2O$

5 $N_2 + H_2 \rightarrow NH_3$

» Extend: Mass calculations

Because we know that mass is always conserved in chemical reactions, we can perform calculations to work out the mass of one substance if we know the masses of the other reactants and products.

The reaction between magnesium and oxygen

▲ Magnesium ribbon

▲ Magnesium oxide in the crucible after the reaction has finished

1 Measure 10 cm of magnesium ribbon. Record its mass on a top-pan balance as accurately as you can, ideally to 2 decimal places, which is the nearest 100th of a gram.

2 Measure the mass of an empty and clean crucible, with a lid.

3 Coil the magnesium ribbon and place it into the crucible. Put the lid on top.

4 Heat from underneath very strongly with a hot Bunsen flame. Lift the lid every minute for 10s to allow oxygen to reach the magnesium. The magnesium will burn very brightly when it reacts with the oxygen. Do not look directly at the flame.

5 Heat for 8 minutes and then allow to cool for 2 minutes.

6 Weigh the crucible and the magnesium oxide inside it.

7 Subtract the mass of the empty crucible from step 2 to find the mass of the magnesium oxide.

Here are some typical results:

Mass of magnesium: 0.48 g

Mass of empty crucible and lid: 34.21 g

Mass of crucible, lid and magnesium oxide: 35.01 g

Therefore, mass of magnesium oxide: 35.01 − 34.21 = 0.80 g

In this experiment, it is impossible to measure the mass of the oxygen which reacted. However, you can calculate the mass of oxygen because you know the mass of the magnesium and the mass of the magnesium oxide.

magnesium + oxygen → magnesium oxide

0.48 g + ?? g = 0.80 g

So the mass of the oxygen is given by:

mass of oxygen = mass of magnesium oxide − mass of magnesium

$$= 0.80 - 0.48$$

$$= 0.32 \text{ g}$$

The thermal decomposition of copper carbonate

Copper carbonate ($CuCO_3$) decomposes when heated to produce copper carbonate (CuO) and carbon dioxide.

Tasks

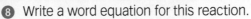

8 Write a word equation for this reaction.
9 Write a symbol equation for this reaction.
10 Explain why the mass decreases in this reaction.

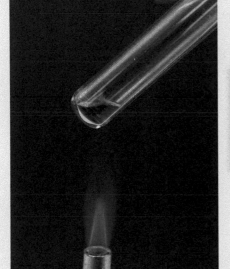

▲ The thermal decomposition of copper carbonate

A student weighed an empty boiling tube. It had a mass of 41.24 g. She then weighed 1.24 g of copper carbonate into the boiling tube and heated it strongly for 5 minutes. The boiling tube and the copper oxide had a total mass of 42.04 g.

Tasks

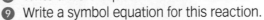

11 Calculate the mass of copper oxide produced in this reaction by subtracting the mass of the boiling tube from the total mass at the end of the reaction.
12 Write the mass of the copper carbonate and the mass of the copper oxide underneath the relevant formula in the symbol equation
13 Calculate the mass of carbon dioxide that was produced in this reaction.

Enquiry:
Burning alcohols

▲ An expedition stove that uses ethanol as a fuel

Alcohols are a family of compounds which are made from carbon, hydrogen and oxygen atoms. Alcohols are often used as fuels. This is because of a number of factors:

- They are liquids, so are easy to transport and store.
- They burn easily.
- They release lots of energy.

There are lots of different alcohols, but they all have names that end in -ol. The alcohol with the smallest molecules is methanol. The most well-known alcohol is called ethanol. Ethanol is found in alcoholic drinks. Ethanol is less poisonous than the other alcohols, but it is still harmful to humans in large quantities. Ethanol is often sold with the name 'methylated spirits' to be used as a fuel and a solvent. Some expedition stoves like the one on the left use methylated spirits (ethanol) as a fuel.

You can compare the energy released by different alcohols using an investigation. Here is a photograph of the apparatus you could use.

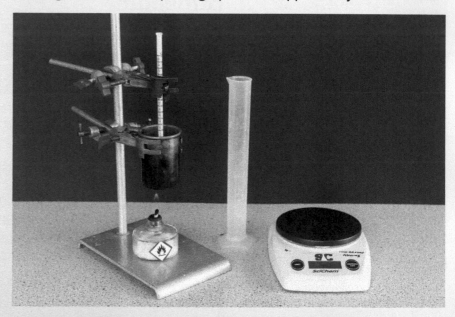

▲ Apparatus used to measure the energy released when a fuel burns

Method

1 A measured amount (e.g. 1g) of one type of alcohol is placed into the alcohol burner.

2 100 cm³ of water is placed into the copper beaker.

3 The starting temperature of the water is measured and recorded.

4 The burner is placed under the copper beaker and lit.

5 After all the fuel has burned, the temperature of the water is recorded.

6 The water in the copper beaker is replaced and the experiment is repeated using a different alcohol.

① What variables should be controlled in this investigation in order to make it a fair test?

② What is the independent variable in this investigation?

③ What is the dependent variable?

④ Fill in the risk assessment table below to identify the main hazards in this investigation and the safety precautions which should be taken to reduce the risks to an acceptable level.

Hazard	Risk (consequence)	Precautions

⑤ One of the big sources of error in this experiment is heat loss, so not all the energy released by the fuel burning is transferred. Looking at the photograph of the apparatus, suggest what changes could be made to the method to reduce the heat loss.

⑥ Looking at the symbol equation below, explain why the mass of the spirit burner decreases during the experiment.
$$C_2H_5OH(l) + 3O_2(g) \rightarrow 2CO_2(g) + 3H_2O(g)$$

Results

Name of alcohol	Temperature rise of water in copper beaker (°C)			
	1	2	3	Mean
Methanol	11	14	11	12
Ethanol	18	20	22	
Propanol	28	28	28	28
Butanol	35	37	9	27
Pentanol	46	44	42	44

⑦ Calculate the mean temperature rise for ethanol.

⑧ Why is a bar chart the most appropriate type of graph to plot for the mean temperature rise of these alcohols?

⑨ Plot a bar chart of the mean temperature rise against the name of the alcohol.

⑩ Which alcohol has the smallest range?

⑪ The alcohols have been listed in the table in order of molecule size, with the smallest molecule at the top (methanol) and the largest alcohol molecule of those tested at the bottom (hexanol). Secondary research suggested that there should be a clear trend: the larger the alcohol molecule, the more energy is released. Which result does not fit this trend?

⑫ Looking at the results table, explain why the mean temperature rise for this alcohol is an anomaly.

⑬ Suggest what could be done in order to get a more accurate result to plot on the graph.

145

Earth

Learning objectives

13 Climate

In this chapter you will learn...

Knowledge

- how to describe processes that increase and decrease the amount of carbon dioxide in the atmosphere
- how to describe the carbon cycle
- how to describe how the greenhouse effect works and name some greenhouse gases
- how to describe the composition of the Earth's atmosphere
- the definitions of the terms global warming, fossil fuels, carbon sink and greenhouse effect

Application

- how to explain how human activities affect the carbon cycle
- how to explain how global warming can affect climate and local weather patterns

Extension

- how to evaluate the implications of a proposal to reduce carbon emissions
- how to evaluate claims that human activity is causing global warming or climate change
- how to compare the relative effects of human-caused and natural global warming

14 Earth resources

In this chapter you will learn...

Knowledge

- how to describe the issues associated with extracting limited natural resources from the Earth
- that recycling reduces the need to extract resources
- that the more reactive a metal, the more difficult it is to separate it from its ore
- that carbon displaces less reactive metals, while electrolysis is needed for more reactive metals
- the definitions of the terms natural resource, mineral, ore, extraction, recycling and electrolysis

Application

- how to explain why recycling of some materials is particularly important
- how to describe how Earth's resources are turned into useful materials or recycled
- how to justify the choice of extraction method for a metal, given data about reactivity
- how to suggest factors to take into account when deciding whether extraction of a metal is practical

Extension

- how to suggest ways in which changes in behaviour and the use of alternative materials may limit the consumption of natural resources
- how to suggest ways in which waste products from industrial processes could be reduced
- how to use data to evaluate proposals for recycling materials

13 Climate

Your knowledge objectives

In this chapter you will learn:
- how to describe processes that increase and decrease the amount of carbon dioxide in the atmosphere
- how to describe the carbon cycle
- how to describe how the greenhouse effect works and name some greenhouse gases
- how to describe the composition of the Earth's atmosphere
- the definitions of the terms global warming, fossil fuels, carbon sink and greenhouse effect

See page 147 for the full learning objectives.

≫ Transition: The water cycle and the carbon cycle

The water cycle describes how water moves through different places in the world, and how it changes state during this journey. You learned about the water cycle in Pupil's Book 1, Chapter 9.

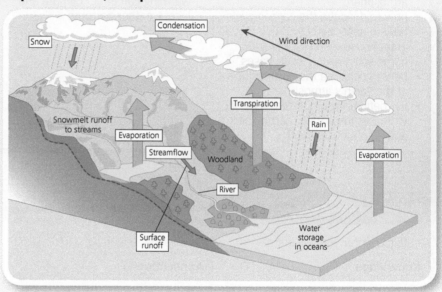

▲ The water cycle

Many of the processes in the water cycle are useful to humans. Farmers rely on rain to help crops grow. As rivers flow downhill to the oceans, they are used for transportation, power and enjoyment.

The carbon cycle

Carbon also moves through a cycle on Earth. This cycle is called the carbon cycle. A simplified version of the carbon cycle can be described like this:

- Carbon atoms are removed from the air by plants during photosynthesis.

- The plants are then eaten by animals, so the carbon atoms are passed into animals.

- Animals respire, releasing energy from glucose in their cells. They release carbon atoms (as CO_2) back into the air through their breath.

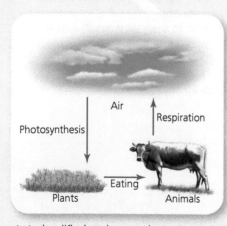

▲ A simplified carbon cycle

Photosynthesis and respiration

You will learn about photosynthesis in Chapter 18. The word equation for photosynthesis is:

carbon dioxide + water → glucose + oxygen

Photosynthesis takes place inside plants when sunlight is absorbed by a green pigment called chlorophyll.

Respiration is the chemical process that takes place inside all living cells. It is how organisms release energy from glucose. You will learn about respiration in Chapter 17. The word equation for respiration is:

glucose + oxygen → carbon dioxide + water

All plant cells respire as well as animal cells. This means that plants release some carbon dioxide into the air. But they absorb a lot more carbon dioxide than this when they photosynthesise during the day.

You can see from the two equations that (chemically speaking) the processes of respiration and respiration are opposite.

Worked example

Describe how carbon atoms are removed from the air in the carbon cycle. Use a word equation in your answer.

Carbon atoms are removed from the air during photosynthesis. The equation for photosynthesis is:
carbon dioxide + water → glucose + oxygen

Apply »

1 Name one of the processes which allows water to move from the atmosphere to the surface of the Earth.

2 How do carbon atoms move from plants to animals?

3 Describe two ways in which humans make use of the water cycle.

4 What is the word equation for respiration?

» Core: The carbon cycle

The atmosphere is mainly made from nitrogen (78%) and oxygen (21%). The remaining 1% is mainly made from argon, but contains a small amount of carbon dioxide.

Carbon is one of the essential elements of life, and is therefore found in all living things. Carbon atoms move through a complicated cycle, including living organisms and non-living things such as rocks and the air. This is called the carbon cycle.

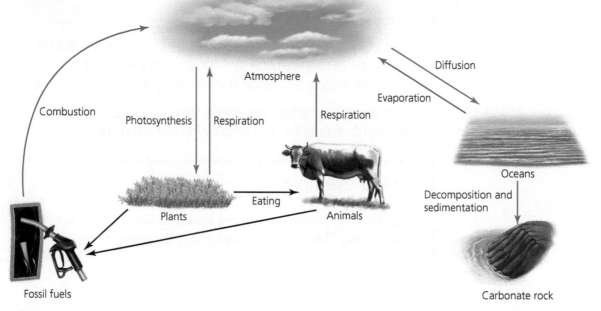

▲ The carbon cycle

Key words

A **carbon sink** is an area of vegetation, the ocean or the soil, which can absorb and store carbon.

Fossil fuels are non-renewable energy resources formed from the remains of ancient plants or animals. Examples are coal, crude oil and natural gas. When a fossil fuel is burned, carbon dioxide is released into the atmosphere.

Removing carbon from the atmosphere

In the diagram, processes which decrease the amount of carbon in the atmosphere are shown with green arrows. This includes photosynthesis, as well as carbon dioxide dissolving in sea water. The oceans are described as a **carbon sink**, because they contain huge amounts of carbon, but can release it again if the sea temperature rises (like when fizzy drinks lose their carbon dioxide fizz and go flat as they warm up). Most of the processes which trap carbon in carbon sinks are very slow, like the formation of carbonate rocks and the formation of **fossil fuels**, both of which take thousands or millions of years.

Adding carbon to the atmosphere

Processes which increase the amount of carbon in the atmosphere (in the form of carbon dioxide) are shown with red arrows. Some of these are natural processes, such as respiration. But burning fossil fuels adds a lot of carbon dioxide to the atmosphere.

Upsetting the balance

Before the population of humans on Earth began to increase rapidly, around 200 years ago, the processes in the carbon cycle were approximately balanced. This meant that the amount of carbon going into the atmosphere was usually balanced by the amount of carbon being removed from the atmosphere and trapped in carbon sinks.

▲ Coal-fired power stations like this release invisible carbon dioxide as well as the visible water vapour from the cooling towers and smoke from the chimney

However, as the population of humans on Earth increases, some of the processes in the carbon cycle are being affected.

- Burning fossil fuels in power stations and vehicles releases carbon that was trapped millions of years ago.

- Cutting down forests reduces photosynthesis and the trees are often burned, releasing carbon dioxide.

- Increasing numbers of humans and farm animals increases respiration, releasing more carbon dioxide.

Worked example

Describe how deforestation (cutting down trees on a large scale) is affecting the carbon cycle.

Trees remove CO_2 from the atmosphere by photosynthesis, so cutting them down causes an increase in CO_2 in the atmosphere. Also, the trees are often burned after they have been cut down, and this releases extra CO_2.

Know >

1 Name two natural processes that remove carbon dioxide from the atmosphere.

2 Name one natural process that releases carbon dioxide into the atmosphere.

3 Name two human processes that increase the amount of carbon dioxide in the atmosphere.

Apply >>

4 Describe some of the effects that the increasing world population is having on the carbon cycle.

Extend >>>

The world population in 1800 was probably less than 1 billion. It is now estimated to be approximately 7.5 billion.

5 Why is it impossible to have exact numbers of the world population for the year 1800?

6 Why is it impossible to have an exact number for the world population today?

7 Algae are simple plants that usually live in water. Many are unicellular, but seaweeds are also algae. Some scientists suggest that increasing the number of algae in the oceans will help to reduce the impact that humans are having on the carbon cycle. Evaluate this suggestion (i.e. consider the pros and cons).

▲ A greenhouse traps energy from the Sun

Key word

The **greenhouse effect** is when energy from the Sun is transferred into the thermal energy store of gases in the atmosphere.

» Core: The greenhouse effect and global warming

Have you ever been inside a greenhouse? Greenhouses are warm inside because they trap energy from the Sun. Our atmosphere also does this, but using gases instead of sheets of glass.

The greenhouse effect

The **greenhouse effect** is a way to describe and explain how the Earth's atmosphere stays warm, trapping energy from the Sun.

1 Waves from the Sun reach the Earth after travelling through space for around 8 minutes.

2 Energy transferred by these waves is absorbed by the surface of the Earth, when it moves into the Earth's thermal energy store.

3 The Earth radiates all of this energy back into the atmosphere, and lots of it passes through the atmosphere and into space.

4 Some of the energy from the Earth is absorbed by gases in the atmosphere, and this warms up the atmosphere.

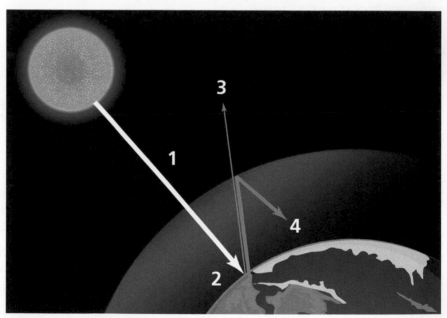

▲ A diagram to explain the greenhouse effect

The gases in the atmosphere which cause the greenhouse effect are mainly water vapour, carbon dioxide and methane. The greenhouse effect is essential for life on Earth, keeping the temperature warm and stable.

Global warming

Scientists know that the average temperature of the atmosphere and the oceans is rising. This is called **global warming**, or climate

Key word

Global warming is the gradual increase in surface temperature of the Earth.

change. Averaged across the whole world, 2015 was the warmest year since records began, and 2016 was even warmer still. However, climate scientists are interested in studying trends over long periods of time (e.g. decades) so if 2016 was not warmer than 2015, this would not mean that global warming had stopped.

Almost all scientists agree that global warming is caused by human activity that is increasing the amount of carbon dioxide and methane in the atmosphere. A few scientists, and some politicians, disagree. But there is convincing evidence that activities that increase the amount of carbon dioxide and methane in the atmosphere are causing global warming.

As global temperatures continue to rise, scientists have made a number of predictions, most of which are already being observed.

- Ice caps in the Arctic and Antarctic will melt and sea levels will therefore rise, flooding low-lying islands.

- Unpredictable weather, such as storms, will be more frequent.

- Droughts will affect farming and may cause famine.

Common error

Some people confuse climate and weather. Weather is a description of the local temperature, wind and rainfall over a short period of time, like a day, week or month. Climate is a description of those factors over a much larger time period, typically a few decades.

Key fact

Don't confuse the greenhouse effect with global warming!

Worked example

Describe how gases in the atmosphere trap energy from the Sun in the greenhouse effect.

Radiation reaches Earth from the Sun. The Earth radiates this energy and most of it passes back into space. Some of the energy radiated by the Earth is absorbed by greenhouse gases in the atmosphere

Know >

1 Name three greenhouse gases.

2 Describe why the greenhouse effect is a good thing for life on Earth.

3 What is meant by the term 'global warming'?

Apply >>

4 Describe some of the consequences of the rise in sea level for people living on islands.

5 'I'm looking forward to global warming because it will make the weather warmer and my holidays more enjoyable.' Give a scientific response to this statement.

Extend >>>

6 'Fifteen of the top 16 warmest years have occurred during the period 2000 to 2016.' Does this statement support the consensus of most scientists about global warming being caused by human activity? Explain your answer.

» Extend: Isn't global warming normal?

The vast majority of scientists agree that global warming is being caused by human activity. But there have been periods of warming and periods of cooling during the past few thousand years. How much evidence do we have that the current rise isn't just caused by normal climate cycles of warming and cooling?

Everyone has heard the term 'ice age'. This is a period of hundreds of years when the global climate is much colder. Humans survived the most recent ice age. Before humans had evolved, there were other ice ages that other species survived. In fact, some experts say that we are still in an ice age at the moment because there are ice sheets in the Arctic and Antarctic. For most of the history of the Earth, scientists think that there was no ice, even at the poles!

How do scientists know?

Scientists have been recording temperatures at many locations around the world for about 150 years. But this doesn't tell us about the temperature of the Earth thousands of years ago. Scientists must rely on indirect evidence and then make estimations of the temperature.

Geologists study rocks and landscapes and see evidence left behind by glaciers. Glaciers are sheets of ice which move slowly downhill, carving huge valleys and leaving behind massive boulders. Glacial landmarks exist in countries that are now too warm for glaciers, suggesting that they used to be much colder.

▲ Glaciers gouge huge valleys out of the landscape

Chemists study the atoms present in sedimentary rocks. For the most recent ice age, bubbles of gas can be found trapped in deep ice, and the gases in these bubbles can also be analysed.

Palaeontologists study the fossil record at different points on the Earth's surface and can make deductions about the climate from the kinds of animals and plants which were alive in a particular period of time.

What do the records show?

Here is an estimate of how the surface temperature of the Earth has changed over the past 450 000 years. It is worth remembering that dinosaurs were alive between 240 and 66 million years ago, so all of the time period represented by this graph is after that period of prehistory. The earliest human-like hominids evolved approximately 2 million years ago, before this graph begins. But stone-age humans with language, cave painting and culture evolved around 50 000 years ago, towards the right-hand side of this graph.

▲ Palaeontologists use fossils to help to estimate what the climate was like millions of years ago

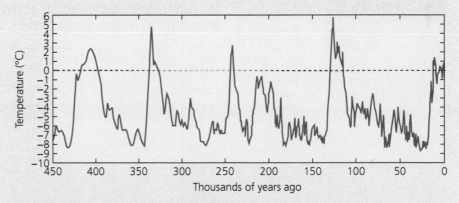

▲ How global temperatures have changed over the past 450 000 years

Tasks

1. How does this graph support the idea that global warming is natural and caused by non-human factors?
2. Suggest how many periods during the past 450 million years there have been with no ice on Earth at all.
3. Why do almost all scientists agree that global warming is taking place and is caused by human activity, rather than natural cycles of ice ages?

Enquiry:
The global warming debate

You learned in Chapter 13 that global warming is definitely happening. No one disputes this because accurate scientific measurements over the past 150 years show a clear trend. However, the causes of global warming are still debated by some people.

The vast majority of people, including almost all scientists, believe that there is enough evidence to be certain that human activity is causing global warming. However, some people disagree, and think that global warming is caused by natural factors. You are going to learn about the evidence and make up your own mind.

Correlation and causation

When we look at a graph showing the average global temperature since 1900 and the total annual amount of carbon dioxide released by human activity for the same period, we see that they have the same trend.

▲ Many people believe that recent unpredictable and extreme weather is being caused by global warming

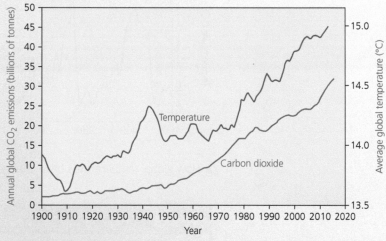

▲ A graph showing the correlation between global CO_2 levels and average temperature

The blue line on the graph shows the carbon dioxide emissions and the scale is shown on the left-hand y-axis. The red line shows the temperature and the scale is shown on the right-hand y-axis. A graph with two y-axes allows us to plot and compare two dependent variables with different units and ranges.

It is important to realise that this graph does not provide evidence that human emissions of CO_2 are causing global warming. But it does show that there is a correlation between the two measurements. This means that they show the same trend.

Correlations like this can provide evidence to support the hypothesis that human CO_2 emissions are causing global warming, but they cannot prove the link.

Key words

A **correlation** is a relationship between two variables. The dependent variable increases or decreases as the independent variable increases.

Evidence is an observation or measurement which supports a hypothesis.

A **hypothesis** is a prediction which can be tested by experiments or observations.

A scientific explanation

In order for scientists and the general public to be convinced that human CO_2 emissions are causing global warming, we need a scientific explanation about how carbon dioxide in the air actually makes the air get hotter.

Scientists have used advanced techniques to understand that the kind of infrared radiation given off by the Earth's surface is absorbed by the actual chemical bonds in every CO_2 molecule. This makes every molecule vibrate, and this means that the energy has been transferred to the thermal store of the carbon dioxide in the air. The air has heated up.

Evaluating secondary sources

When you are trying to make up your own mind about something, it is important to do thorough secondary research. You should think carefully about who published the research. Ask yourself these questions:

- Are the authors qualified scientists?
- Was the research published in a peer reviewed journal?
- Does the research agree with other studies?
- Might the author have a bias, perhaps by being paid by a funder who would benefit from a specific conclusion?
- Did the researcher collect enough data in a valid way?

❶ Why might it not be a sensible idea to trust a study on the causes of global warming which was funded by an environmental conservation charity?

❷ Why might it not be a sensible idea to trust a study on global warming published in the North American Journal of Crude Oil Companies?

Wikipedia™ is an online encyclopaedia which can be added to and edited by anyone who sets up a login account. It is free to use, and approximately 400 million people use it every single month (similar to the entire population of the USA). You do not have to be an expert to edit it, but experts do view and edit pages, so that the information on Wikipedia is usually correct and has a balanced point of view.

❸ Do you think Wikipedia is a good source of secondary information for you as a science student? What are the pros and cons? Describe when you would and would not use Wikipedia.

❹ Write a letter to an imaginary politician who does not believe that humans are causing climate change, explaining the scientific evidence and the probable consequences of allowing CO_2 emissions to continue to rise.

Key words

Secondary data are results which were collected by someone else. Secondary research may also relate to conclusions made by someone else.

A journal is a magazine which publishes science research. Almost all journals are peer reviewed, which means the articles are checked by other expert scientists before being printed.

Bias is when a researcher controls the outcome of an investigation, or when a journal favours a particular point of view.

A funder is a person or organisation which pays for scientific research to be done.

WIKIPEDIA
The Free Encyclopedia

14 Earth resources

Your knowledge objectives

In this chapter you will learn:
- how to describe the issues associated with extracting limited natural resources from the Earth
- that recycling reduces the need to extract resources
- that the more reactive a metal, the more difficult it is to separate it from its ore
- that carbon displaces less reactive metals, while electrolysis is needed for more reactive metals
- the definitions of the terms natural resource, mineral, ore, extraction, recycling and electrolysis

See page 147 for the full learning objectives.

» Transition: Reactivity series

Do you remember the reactivity series, from Pupil's Book 1, Chapter 11? You learned that the most reactive metals are put at the top of the reactivity series and the least reactive metals are placed at the bottom.

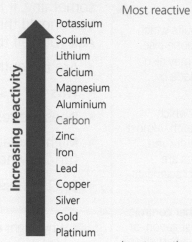

Most reactive

Potassium
Sodium
Lithium
Calcium
Magnesium
Aluminium
Carbon
Zinc
Iron
Lead
Copper
Silver
Gold
Platinum

Increasing reactivity

Least reactive

You could add any metal into the reactivity series if you do experiments to compare how reactive it is compared to other metals. Sometimes you see a reactivity series with just a few metals in it. Sometimes you see a reactivity series with lots and lots of metals in it.

Sometimes it is helpful to put carbon into the reactivity series, even though it is not a metal. It is fun to make up your own mnemonic to help you remember the order of elements in the reactivity series. Here is one: people say little children make a colourful zebra ill like coughing sneezing grunting puking!

Displacement reactions

In a displacement reaction, a more reactive element will replace a less reactive element in a compound. We can make predictions using the reactivity series about whether one element will displace another element in a compound.

▲ In this displacement reaction, copper is displacing silver from silver nitrate solution

If magnesium is mixed with a solution of copper sulfate, the magnesium displaces the copper because magnesium is more reactive than copper:

magnesium + copper sulfate → magnesium sulfate + copper

But magnesium cannot displace sodium from sodium sulfate, because magnesium is less reactive than sodium:

magnesium + sodium sulfate → (no reaction)

Worked example

Predict and explain what will happen if zinc is added to separate solutions of magnesium chloride and copper chloride. Use word equations for any reactions that will occur.

When the zinc is added to the magnesium chloride, there will not be a reaction because the zinc is less reactive than the magnesium, so it cannot displace it. When zinc is added to the copper chloride, a displacement reaction will occur because the zinc is more reactive than the copper:
zinc + copper chloride → zinc chloride + copper

Apply »

1 Name a metal that will displace calcium from calcium nitrate.

2 What would be the products of a reaction between magnesium and lead nitrate?

3 Write a word equation for the reaction between magnesium and iron chloride.

» Core: Finite resources

Resources in the crust

The Earth's crust contains many **natural resources** which humans extract and use. Sometimes we use these resources in their natural form. For example, rock is removed from the ground, crushed into smaller pieces and then used to build roads. But usually, natural resources are processed to make them more useful. For example, chemical reactions are used to **extract** metals from rocks.

Key words

Natural resources are substances from the Earth which act as raw materials for making a variety of products.

To **extract** a metal is to separate it from a compound or mixture.

Elements in the crust

Scientists have estimated the amount of each element present in the Earth's crust. The table below shows the abundance of each element.

Element	Abundance (%)
Oxygen	46
Silicon	27
Aluminium	8
Iron	6
Calcium	5
Magnesium	3
Sodium	2
Potassium	2
Other elements	1

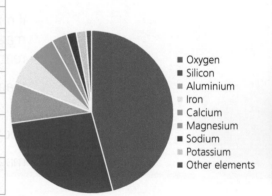

- Oxygen
- Silicon
- Aluminium
- Iron
- Calcium
- Magnesium
- Sodium
- Potassium
- Other elements

Many of these elements are useful. Silicon is used in making electronic devices such as computers and mobile phones. But because silicon is one of the elements in sand (silicon oxide), there is no shortage of it. The challenge is making it pure enough to be used in computer chips.

Aluminium and iron are both very useful metals, but they are not found pure in the ground. They are both found chemically bonded to oxygen, as aluminium oxide and iron oxide.

Finite resources

Most people have heard of the word 'infinity', and the word 'infinite'. Infinite means that something can never run out. We could describe sea water as infinite because it wouldn't really matter how much of it we used, there would still be plenty more. But fewer people know what the word 'finite' means. It is the opposite of infinite. If something is a **finite resource**, then one day it might run out. The faster we remove finite resources from the crust, the sooner they will run out.

Look at the abundance data in the table. You *could* say that all of the elements in the table are finite, but in reality, we are never going to run

Key word

A **finite resource** is a natural resource that might run out in the future if humans are not careful about how quickly it is used up.

out of silicon, aluminium or iron in the crust. Other metal elements have abundances which are too low to be included in the table. For example, indium makes up only 0.000016% of the crust, but it is needed to make the touchscreen on your mobile phone. Indium is certainly a finite resource, which is why it is important to recycle your old phone.

Recycling

Recycling objects when they have reached the end of their usable life is important because it reduces the need to extract resources from the crust. It also saves energy, and it reduces the amount of rubbish thrown away into landfill. We recycle aluminium objects such as drinks cans because it saves energy and reduces landfill, not because we are going to run out of aluminium oxide in the crust.

Key word

Recycling is when an object is processed so that the materials it is made from can be used again.

▲ Aluminium cans which have been crushed, ready for recycling

Worked example

Gemma suggests that recycling magnesium is more important than recycling calcium, but Debra disagrees. Using the data in the table and pie chart, explain why Gemma could be correct.

Magnesium makes up 3% of the Earth's crust and calcium makes up 5%. Magnesium is rarer than calcium, so it is important to reduce the speed at which we extract it from the crust. Recycling magnesium is therefore more important than recycling calcium.

Know ❯

1 Which is the most abundant metal element in the Earth's crust?

2 Which is the most abundant non-metal element in the Earth's crust after oxygen?

3 Give three reasons why recycling is important.

Apply ❯❯

4 Use data from the table or pie chart on the opposite page to suggest why it would be more important to recycle gold than iron.

5 Explain the benefits of recycling aluminium, despite it being so abundant in the crust.

Extend ❯❯❯

6 Glass is made from sand, which is mainly silicon oxide. Using data from the table or pie chart, and your knowledge about recycling, evaluate the pros and cons of recycling glass jars and bottles.

Enquiry

7 Look at the pie chart opposite. Why is a pie chart the best way to represent this data?

8 Why are the percentages in the pie chart estimates and not accurately measured data?

» Core: Extracting metals using displacement reactions

The reactivity series and displacement reactions

Most reactive

Potassium
Sodium
Lithium
Calcium
Magnesium
Aluminium
Carbon
Zinc
Iron
Lead
Copper
Silver
Gold
Platinum

Increasing reactivity

Least reactive

You learned in Pupil's Book 1, Chapter 11, that a displacement reaction occurs between a reactive element and a compound containing a less reactive element. In a displacement reaction, the more reactive element displaces the less reactive element, stealing the other part of the compound. For example, in the thermite reaction:

aluminium + iron oxide → aluminium oxide + iron

The aluminium is more reactive than the iron, so it displaces the iron and 'steals' the oxygen. On the left there is a reminder of the reactivity series. You will see that carbon has been included, even though it is a non-metal.

Metal ores

Key words

A **mineral** is a solid chemical compound found in the Earth's crust. Because each mineral is a compound, it will have a specific chemical formula.

An **ore** is a rock which contains enough of a mineral to make it worth extracting from the crust.

Metals are usually found chemically bonded to oxygen when they are in the Earth's crust. Only a few metals are found pure in the ground. A solid element or compound found in the crust is called a **mineral**. Each mineral has a particular chemical formula. Many minerals contain metal elements bonded to oxygen. Minerals are often mixed together to form rocks. Rocks are impure, so don't have a chemical formula. If a rock contains enough of a mineral to make it worth extracting the rock, we call it an **ore**. Here is an example.

▲ Metal: calcium, Ca

▲ Mineral: calcium carbonate, $CaCO_3$

▲ Rock: limestone

Extracting metals from their compounds

We can use displacement reactions to extract some metals from their oxide compounds. Carbon is a cheap raw material that can be used to displace metals which are lower than carbon in the reactivity series. For example:

iron oxide + carbon → iron + carbon dioxide

copper oxide + carbon → copper + carbon dioxide

The metal oxide must be heated with carbon to a very high temperature. The first metals to be extracted in this way were copper and then iron, which gave rise to the Bronze Age around 5000 years ago and then the Iron Age around 3000 years ago.

Unreactive metals

The metals nearest to the bottom of the reactivity series are so unreactive that they don't normally form compounds with oxygen. This means that they can be found pure in the Earth's crust, mixed in with rock but not chemically bonded to any other elements. To extract the metal, you just need to break apart the rock or melt the metal.

▲ Gold metal is found pure in the middle of this rock

Worked example

Suggest how lead is extracted from lead oxide, which is a compound present in lead ore. Use a word equation in your answer.

Lead ore is mixed with carbon and heated to a high temperature. The reaction is a displacement reaction:
lead oxide + carbon → lead + carbon dioxide

Know >

1 Define the term 'mineral'.

2 Define the term 'ore'.

3 Name a metal which is found pure in the crust.

Apply >>

4 Suggest how platinum is extracted from rock after the platinum ore has been removed from the ground.

5 Pam suggested that magnesium could be extracted from magnesium carbonate by heating it with carbon to a high temperature. Would this work? Explain your answer.

Extend >>>

6 Hydrogen can be placed into the reactivity series, between lead and copper. State and explain which of the following metals could be extracted by heating the ore with hydrogen: aluminium, copper, iron, zinc.

» Core: Extracting metals using electrolysis

Some metals are too reactive to be extracted using a displacement reaction that involves carbon. A more expensive method of extraction is needed, which is called **electrolysis**. Electrolysis is when electricity is used to split a compound into its elements.

Extracting aluminium

Aluminium ore is a reddish-brown rock called bauxite. It contains the mineral aluminium oxide, which is a white powder once the impurities have been removed.

▲ Bauxite ore

▲ Aluminium oxide

To electrolyse the aluminium oxide, it must be added to a hot liquid called cryolite. The temperature of the cryolite is almost 1000 °C! An electric current is then passed through the solution of aluminium oxide in the molten cryolite. Here is a diagram of the way that this is done.

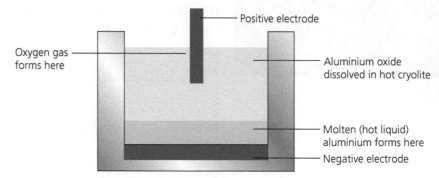

▲ The apparatus used to extract aluminium from aluminium oxide by electrolysis

The aluminium oxide splits up when an electric current is passed through it:

aluminium oxide → aluminium + oxygen

The aluminium sinks to the bottom of the tank and because of the very high temperature, it is a liquid. It is sucked out of the bottom of the tank using a siphon tube. The oxygen gas is produced at the positive electrode at the top of the tank.

The history of electrolysis

Humphry Davy was not the first person to use electrolysis, but he was the first person to use it to extract reactive metals from their compounds. At the time, no one knew that these metal elements even existed! He was so successful using this method, that having discovered potassium in 1807, he then went on to discover sodium, barium, calcium and magnesium the following year.

▲ Sir Humphry Davy, 1778–1829

A summary of extraction methods

You have now seen that the method used to extract a metal from the Earth's crust depends on its reactivity. This is summarised in the diagram below.

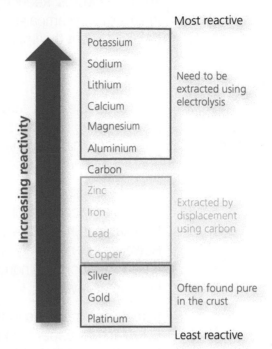

▶ A summary of the extraction methods for different metals

Worked example

Lithium can be made by the electrolysis of lithium bromide. Name the two products of this process, and write a word equation for the reaction.

The two products are lithium and bromine.
The word equation is:
lithium bromide → lithium + bromine

Know >

1 How is aluminium extracted from aluminium oxide?

2 What is the name of the ore which contains aluminium oxide?

3 Who used electrolysis to discover several new metallic elements?

Apply >>

4 One way of extracting magnesium is by electrolysing molten magnesium chloride. Suggest the names of the two products of this process, and write a word equation for the reaction.

5 List the factors that should be taken into account when deciding whether extracting a metal is practical.

Extend >>>

6 Some metals are expensive because they are rare, like platinum and gold. Aluminium is not rare. On page 158 we saw that aluminium is the most abundant metal element in the crust. But aluminium is expensive – much more expensive than iron. Using ideas about how it is extracted, suggest two reasons why aluminium is expensive.

Enquiry >>>>

7 Strontium is a metal which is more reactive than calcium but less reactive than lithium. Suggest a method for extracting strontium from strontium oxide.

8 Tin is a metal which is more reactive than lead but less reactive than iron. Suggest how it is extracted from tin oxide.

» Extend: Reduce, reuse, recycle

The three words 'Reduce, Reuse, Recycle' are often used together in advertising campaigns aimed at getting consumers to think carefully about the impact they have on the environment.

REDUCE REUSE RECYCLE

▲ An advert to encourage people to think about their impact on the environment

Reduce

Making people think carefully about how much of a particular resource they use is important because natural resources are finite. This means that they might run out one day if we are not careful about the speed at which we extract them from the crust.

Even if the resource we are using is not likely to run out, there are often energy costs involved with making it from natural resources. For example, in the UK we are not likely to run out of fresh water. However, lots of energy is needed to purify it to make it safe to drink, and to transport it to homes, so it is important to be careful about how we use water.

Sometimes there are indirect consequences on the environment. For example, paper is made from wood which is grown especially for making paper. But making paper uses quite a lot of chlorine to make it white, and chlorine is very dangerous for living organisms.

▲ Dual flush toilets allow us to choose how much water we use to flush the toilet, helping us to save water at home

▲ The paper bag has a lower impact on the environment because it will rot away in landfill

Alternative materials that have a lower impact on the environment can often be used. Before 2011, plastic carrier bags were given away for free by shops across the UK when customers bought something. Wales introduced a 5p charge for every bag in 2011, Northern Ireland followed in 2013, Scotland in 2014 and England in 2015. Paper bags are not taxed in this way, because they rot away in landfill sites, and are not as damaging for the environment.

Reuse

Sometimes things can be reused instead of being thrown away. If you do have a plastic bag from a shop, can you reuse it next time you go shopping? In recent years, there has been a trend for 'upcycling'. This is when something at the end of its life is changed into something new.

▲ Upcycling has made this shoulder bag from an old pair of jeans

Recycle

Recycling saves natural resources and also helps to conserve energy. This is because it always takes less energy to recycle a material than it does to make it from natural resources.

▲ Some of the items we can recycle from our household waste

But it doesn't always make sense to recycle everything. Fruit juice drinks are sometimes sold in boxes made from cardboard, aluminium foil and plastic, stuck together in a way that makes them difficult to recycle.

Recycling cartons like these leaves you with cardboard pulp which can be made into new cardboard products. But separating the aluminium foil from the plastic is often too difficult, so instead the mixture of plastic and foil is put into building materials and children's play equipment. This is certainly better than the carton ending up in landfill, but probably not as good as being able to recycle all of the parts of the carton. In other parts of the world, the complicated machines needed to recycle these cartons may not be available, so the cartons will end up in landfill.

▲ Cartons like this are difficult to recycle

Tasks

1. Design an infographic poster to encourage people to think carefully about how their lifestyle choices impact on the use of energy and natural resources.
2. Write a letter to your local council or MP to encourage an improvement in the amount of waste that is recycled where you live. You can usually find out the percentage of local waste that is recycled by visiting the local recycling centre.

Enquiry:
Is fracking a good idea?

▲ Diesel and petrol are both fossil fuels obtained from crude oil

Amongst the most valuable and important natural resources in the crust are the **fossil fuels**. These are coal, crude oil and natural gas.

Crude oil is processed into a range of products, many of which are fuels. These include petrol, diesel, central heating oil and aeroplane fuel. But chemicals obtained from crude oil are the starting point for most man-made fibres used in clothing, and almost all plastics.

How much oil is left?

Crude oil is a finite resource, and we are extracting it from the crust very quickly. This means that one day there will be so little left that we will need to find alternatives. Crude oil will have become so hard to find and extract that it will be too expensive for people to buy.

Scientists aren't sure how many years are left until we cannot extract any more crude oil from the crust. Politicians, business leaders and consumers ask scientists to estimate how many years are left until we find it too difficult and expensive to extract oil.

> **Key word**
>
> An **estimate** is when you make an educated guess at a value, using scientific ideas, data or a rough calculation to help you to decide on your chosen value.

What is fracking?

Hydraulic fracturing (called 'fracking') is a way of extracting fossil fuels from rocks deep underground. These rocks contain some fossil fuels, but not enough to use a normal drilling technique to make a standard oil well. So fracking allows engineers to remove fuels from the crust that could not be removed in other ways.

In fracking, a hole is drilled deep into the rocks, perhaps more than 2000 m below the surface. Horizontal holes are often drilled outwards from this vertical well. A mixture of water and sand (and some other chemicals) are then injected into the well, to break apart the rocks and allow the oil or gas to be released into the well after the water supply is shut off. The oil or gas can then be pumped to the surface.

◄ A simplified diagram (not to scale) showing how fracking works. The blue arrows represent the liquid mixture which is injected into the rock. The orange arrows represent the fossil fuel

Why are people opposed to fracking?

Some people do not like the idea of fracking because they might have chosen to live in a quiet and beautiful part of the countryside and they do not want increased traffic, buildings and noise.

There is evidence that fracking causes mini-earthquakes, although the tremors are normally too weak to be felt at the surface. In some countries, chemicals are added to the water/sand mixture which are toxic, but in the UK this would not be allowed. However, some people still have concerns that fracking might cause water pollution. A report published in 2012 by the Royal Society and the Royal Academy of Engineering (both scientific organisations) concluded that the environmental risks from fracking were low.

Some people believe that instead of developing fracking as a way to extract more fossil fuels from the crust, we should instead focus on finding alternative energy resources which do not cause global warming.

▲ People protesting against a proposed fracking plant in the UK

▲ Offshore wind farms like this one off the coast of Essex could help to provide more energy for the UK

❶ Many people talk about how long it will take until we 'run out' of crude oil in the Earth's crust. In fact, we will never extract *all* of the crude oil in the crust. Suggest why this is.

❷ Suggest reasons why scientists do not know how long it will be until crude oil is too difficult and expensive to extract from the crust.

❸ Who would benefit from fracking in the UK, and how would they benefit?

❹ Some people are worried that surface water (rivers and lakes) could become polluted due to fracking. What effect would this have on ecosystems?

❺ In the USA, fossil fuels are owned by the person who owns the land above the fossil fuel reserve. In the UK, fossil fuels are owned by the government. How would this affect who benefits from fracking?

❻ Produce a report for the government on whether or not fracking should be allowed in your local area. Justify your decision using factual evidence which you have researched from reliable secondary sources.

Organisms

Learning objectives

15 Breathing

In this chapter you will learn ...

Knowledge

- that oxygen and carbon dioxide move between alveoli and the blood in gas exchange
- that cells require oxygen for aerobic respiration and carbon dioxide is a waste product
- how breathing occurs
- that the rate of breathing is determined by the amount of oxygen the body needs
- the definitions of the terms trachea, bronchi, alveoli, ribs, diaphragm and lung volume

Application

- how to explain how exercise, smoking and asthma affect the gas exchange system
- how to explain how the parts of the gas exchange system are adapted to their function
- how to explain changes in breathing rate and volume during exercise and at rest
- how to explain how changes in volume and pressure inside the chest move gases in and out of the lungs

Extension

- how to evaluate a possible treatment for a lung disease
- how to predict how a change in the gas exchange system could affect other processes in the body
- how to evaluate a model for showing the mechanism of breathing

16 Digestion

In this chapter you will learn ...

Knowledge

- what a balanced diet is and why it is important
- that the function of the digestive system is to break large food molecules into smaller ones
- that iron is important for red blood cells and calcium is needed for strong teeth and bones
- that vitamins and minerals are needed in small amounts to keep the body healthy

Application

- how to suggest the possible health effects of an unbalanced diet
- how to suggest the food requirements for a healthy diet, using information provided
- how to describe the adaptations of the parts of the digestive system
- how to describe the process of digestion

Extension

- how to suggest a diet for a person with specific dietary needs
- how to analyse nutritional information from food packaging
- how to explain how specific symptoms can indicate problems with the digestive system

15 Breathing

Your knowledge objectives

In this chapter you will learn:
- that oxygen and carbon dioxide move between alveoli and the blood in gas exchange
- that cells require oxygen for aerobic respiration and carbon dioxide is a waste product
- how breathing occurs
- that the rate of breathing is determined by the amount of oxygen the body needs
- the definitions of the terms trachea, bronchi, alveoli, ribs, diaphragm and lung volume

See page 171 for the full learning objectives.

» Transition: What makes up an organism?

Have you ever wondered what makes up a human body? What exactly are you made from? Well you have about 35 million billion cells. Each of these cells contain water, and so this makes up about 65% of you. Of what is left, the average person is about 20% protein and about 10% fat. However, this can vary depending upon your body shape.

In your body, a group of the same type of cells in the same location is a tissue. So muscle cells in the same place are called a muscle tissue. Two or more tissues in the same place makes an organ – so your heart is made from muscle and nerve tissues. Organ systems are made from two or more organs not necessarily in the same place but doing the same thing. Your circulatory system is made from your heart and blood vessels, which circulate your blood.

You are made from several more important organ systems. Your digestive system breaks down food. You will learn more about this in Chapter 16. Your respiratory system is made of your lungs and airways and helps get oxygen into your blood. You will learn more about this later in this chapter. Your nervous system is made from your brain, spinal cord and nerves that run all over your body.

Your nervous system transfers information from your sense organs to your brain. This information travels as an electrical signal along nerves that spread out throughout your entire body. Can you name all your senses and the organs that work with them? You see with your eyes. You smell with your nose. You taste with your tongue. You hear with your ears. You touch with your skin. We call these 'sense organs'.

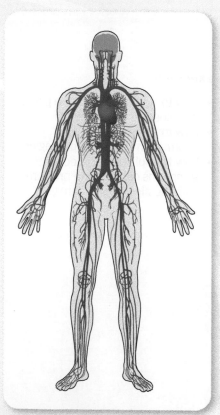

▲ Your circulatory system

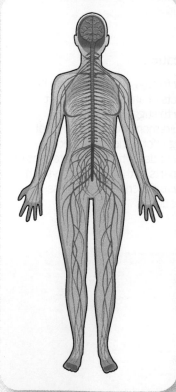

▲ Your nervous system

Many doctors now think that we have more than five senses. As well as touch, your skin can detect the amount of pressure an object touches you with, and the pain it might cause. Amazingly our ears help us balance as well as hear. Balance is another of your senses.

Your skeletal system is all of the bones in your body and your muscle system is all of your muscles. These contract and relax to help you move. You learned about how you move using your muscles and bones in Pupil's Book 1, Chapter 15.

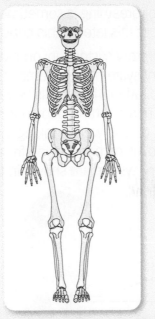 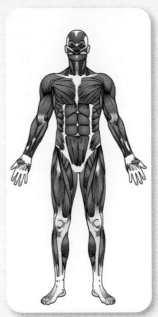

▲ Your skeletal and muscular systems

Worked example

What effect would damage to the lungs have?

Damage to the lungs would stop oxygen getting into the blood so easily. This would mean that less respiration could take place in the cells and the affected person would have less energy and feel tired.

Know >

1 What percentage of you is made from water?
2 What is a tissue?
3 What two types of tissue make up your heart?
4 What are your five senses?
5 What other senses do you have?

Apply >>

6 What would happen if all the bones or muscles in your body were removed?

Key words

The **respiratory system** replaces oxygen and removes carbon dioxide from blood.

Breathing is the movement of air in and out of the lungs.

The **trachea** (windpipe) carries air from the mouth and nose to the lungs.

» Core: The respiratory system

The **respiratory system** could be called the breathing system. This would remind us that it is not to do with respiration but the process of **breathing**. Respiration is the chemical reaction which releases energy from glucose. This constantly happens in all your body's cells and provides them with the energy they need to live. The process of breathing is sometimes called ventilation. You will learn more about this later in this chapter.

The respiratory system absorbs oxygen into your blood. Your circulatory system them pumps this to all of your cells, which need it for respiration. These cells release carbon dioxide as a waste product. This is returned to your lungs in your blood and then breathed out. As well as taking in oxygen, your respiratory system also removes carbon dioxide from your body. The key parts of the respiratory system are shown in the photo below.

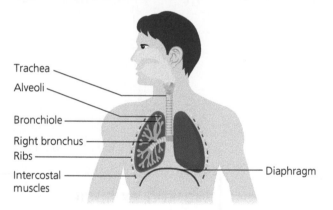

▲ The parts of the human respiratory system

Trachea

Your **trachea** is a long tube that runs from the back of your mouth to just above your lungs. It has a rigid structure and stays open day and night to allow you to breathe. A second tube, called the oesophagus, runs from the back of your mouth to take food to your stomach. You will learn more about this in Chapter 16.

If you run your fingers up and down the front of the lower part of your neck you might be able to feel your trachea. You might be able to feel ridges along its edge. These are made of cartilage, which is the same substance that makes your nose and ears. These rings of cartilage keep your trachea rigid and open at all times.

The lining of the trachea is covered with cells with tiny hairs. These are called ciliated cells and the hairs are called cilia. These beat in a rhythmical motion like a Mexican wave. They move any mucus and dust or pathogens upwards to stop them entering your lungs. You swallow this mucus without realising it.

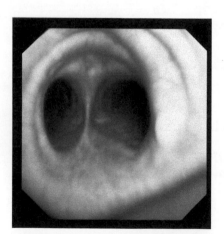

▲ The trachea as it splits into bronchi

▲ Tiny hairs called cilia line the trachea

Bronchi and bronchioles

Key words

The **bronchi** are two tubes which carry air to the lungs.

Bronchioles are small tubes in the lung.

Just above your lungs your trachea splits into two **bronchi**. One travels to each lung. We say two bronchi and one bronchus. So as each bronchus enters a lung it splits into thousands of smaller branches called **bronchioles**. These branch again and again throughout your lungs.

Alveoli

At the end of all bronchioles are **alveoli**. These are sometimes called air sacs. They resemble a bunch of blown-up balloons. The transfer of gases to and from the blood happens in alveoli. You will learn more about this on the next section. They have a huge surface area to maximise the movement of these gases.

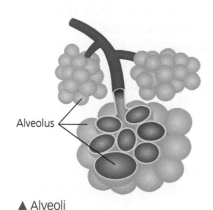

Alveolus

▲ Alveoli

Ribs

Your **ribs** are bones that form a cage around your vital organs in your chest, including your heart and lungs. Between your ribs are intercostal muscles. If you have eaten the meat from spare ribs in a restaurant, you have eaten intercostal muscles! These contract and relax when you breathe in and out. You will learn more about the process of breathing later in this chapter.

Diaphragm

Your **diaphragm** is a sheet of muscle that sits below your lungs. Like your intercostal muscles, the diaphragm contracts and relaxes when you breathe.

Key words

Alveoli are the small air sacs found at the end of each bronchiole.

The **ribs** are bones that surround the lungs to form the ribcage.

The **diaphragm** is a sheet of muscle found underneath the lungs.

Know ❯

1 What does the respiratory system do?

2 What two parts of your body does your trachea connect?

3 What muscles are found between your ribs?

4 What adaptation keeps your trachea open?

5 How are ciliated cells adapted?

Apply ❯❯

6 Why do bronchioles have smaller diameters than bronchi or tracheas?

7 Describe the journey of an oxygen molecule as it moves from the air to inside your blood.

Extend ❯❯❯

8 What would happen if all the capillaries were removed from your lungs?

9 What are the similarities and differences between fish gills and human lungs?

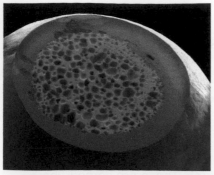

▲ A section through a mint aero bubble chocolate. Is this a good model for the structure of alveoli?

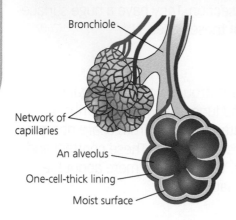

Bronchiole

Network of capillaries

An alveolus

One-cell-thick lining

Moist surface

▲ A cluster of alveoli showing their key adaptations

» Core: Alveoli in detail

Your lungs are amazing organs. They allow you to breathe air in and out. When you breathe in, oxygen moves from the air to your blood. When you breathe out, carbon dioxide moves from your blood to the air, which you then breathe out.

Each of your lungs has millions of tiny alveoli or air sacs. These look like the tiny bubbles in a bar of bubbly chocolate. It is in these tiny alveoli that gases are exchanged between your blood and the air. Because there are so many alveoli in each lung, the total surface over which gases can diffuse is huge. It is approximately half the size of a tennis court. This is a key adaptation of your lungs.

Alveoli are surrounded by a rich blood supply. Millions of tiny capillaries carry blood through the lungs, picking up oxygen and depositing carbon dioxide. This adaptation means your lungs can maximise the transfer of gases.

The lining of the alveoli is very thin (usually only one cell thick) and moist. This means that gases can move faster though the membranes of the alveoli into and out from the blood.

What happens in an alveolus?

Diffusion is the movement of substances from an area of high to lower concentration. Gases like your deodorant diffuse. They spread out from a high concentration (perhaps under your arms if you have just sprayed yourself) to a lower concentration everywhere else. Liquids also diffuse. When you put a teabag in a cup of water, the tea diffuses from a high concentration inside the teabag into the water outside. This happens naturally.

When you breathe in, air containing lots of oxygen moves into your lungs. Here the oxygen is at a high concentration in your alveoli and in a low concentration in your blood. So it moves by diffusion from inside your alveoli into your blood. Because your blood is always moving, the blood that is now high in oxygen is pumped to the rest of your body. The blood is replaced with that which has come from your body and so is low in oxygen. This way oxygen is always moving from your alveoli into your blood. Carbon dioxide moves in the other direction.

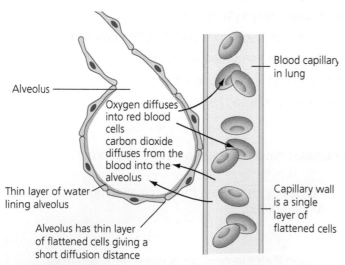

Blood capillary in lung

Alveolus

Oxygen diffuses into red blood cells carbon dioxide diffuses from the blood into the alveolus

Thin layer of water lining alveolus

Alveolus has thin layer of flattened cells giving a short diffusion distance

Capillary wall is a single layer of flattened cells

▲ Diffusion of gases from the alveoli to the blood

Smoking, exercise and asthma

We have known for many years that smoking tobacco is bad for your health. There are over 4000 chemicals in cigarettes and 43 of them are known to cause cancer. People continue to smoke despite knowing the dangers because nicotine in tobacco is addictive. Another chemical in cigarettes called tar causes cancer. It is a carcinogen.

Lung cancer is a difficult disease to treat. As with many cancers, it is important to discover, diagnose and start treatment as early as possible. Lung cancer can be treated by chemotherapy and radiotherapy. These use drugs and radiation respectively, to kill the cancerous tumour. It is difficult to kill just the tumour cells however, and not affect the healthy cells around the tumour. Both are painful treatments which often have side effects, including hair loss, tiredness and fatigue. Another option is an operation which removes the cancer. Again it is hard to remove just the tumour cells. Any major operation like this is difficult for the patient. Recovery is often slow and painful.

Smoking reduces the surface area of your alveoli. The hot gases that are breathed in break down the lining of the alveoli. This reduces the exchange of oxygen and carbon dioxide. Smokers often find it difficult to exercise. This is because they are unable to get enough oxygen to their muscles for respiration to provide energy.

The hot cigarette smoke also kills the ciliated cells that line the airways. This means that smokers are often less able to remove mucus with dirt and pathogens from their airways. This leads to a characteristic smoker's cough.

We've also known for many years that exercise is good for your health. When you exercise you strengthen the intercostal muscles and your diaphragm. This makes the process of breathing easier.

Asthma is a medical condition that many doctors think is both inherited from our parents but also linked to the higher levels of air pollution that we now have. Sadly, there is no cure for asthma. An asthma attack occurs when the lining of the airways become irritated. This means they swell and so become narrower. This makes breathing much more difficult during an attack. People who suffer from asthma can reduce the frequency of their attacks by staying away from air pollution, such as cigarette smoke, camp fires or polluted cities. They can also use asthma inhalers to reduce the swelling, returning breathing to normal.

Know >

1 What gas enters your blood in the alveoli?

2 What gas leaves your blood in the alveoli?

3 By what process do gases enter and leave the blood?

4 How are your alveoli adapted to maximise movement of gases?

Apply >>

5 Why does oxygen diffuse into your blood?

Extend >>>

6 Where else does diffusion happen in your body?

7 How would you investigate the link between air pollution and asthma?

» Core: The process of breathing

The scientific name for breathing is ventilation. Simply put, this process replaces the air in your lungs. At rest, most people breathe about 12 to 20 times a minute. During exercise the rate of breathing can easily double. Also, the depth of each breath increases. At rest, adults will breathe in about half a litre of air. During exercise this can almost double, so your breathing rate depends upon the amount of oxygen your body needs. When you need more your breathing rate and depth increases. These changes are involuntary. You don't choose to breathe more or less frequently; it is automatic.

The total **lung volume** of an adult female and male lung is about 4 and 6 litres, respectively. You can measure this by following the simple experiment described in the Enquiry spread at the end of this chapter.

> **Key word**
>
> The **lung volume** is a measure of the amount of air breathed in or out.

Inhalation

The process of breathing in is called inhalation. When you inhale, the following steps occur:

1 Your intercostal muscles contract, which moves your ribs up and out.

2 Your diaphragm contracts, which moves it downwards.

3 These two steps make the volume of your chest cavity bigger.

4 Air rushes in from outside to fill this extra space.

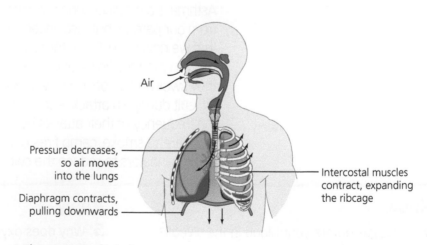

Air

Pressure decreases, so air moves into the lungs

Diaphragm contracts, pulling downwards

Intercostal muscles contract, expanding the ribcage

▲ The process of inhaling

Exhalation

Breathing out is called exhalation. The steps are the reverse of inhalation:

1 Your intercostal muscles relax and so move your ribs downwards.

2 Your diaphragm relaxes, which moves it upwards into a curved position.

3 These two steps make the volume of your chest cavity smaller.

4 Air is expelled from your lungs.

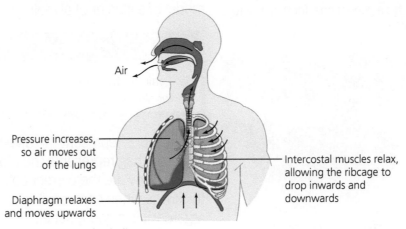

Air

Pressure increases, so air moves out of the lungs

Intercostal muscles relax, allowing the ribcage to drop inwards and downwards

Diaphragm relaxes and moves upwards

▲ The process of exhaling

If you put your hands on your chest as you breathe in, you can feel your ribcage move upwards and outwards. Do this now as you are reading this sentence. Did you suck air into your lungs? Probably not, unless you are reading this as you exercise! Our intercostal muscles and diaphragm increase the volume of our chest. Air is not normally sucked in. Try sucking in air now. Do you see how different that is from breathing normally?

Know >

1 What is the scientific name for breathing?

2 What does exhalation mean?

3 What happens to your intercostal muscles when you breathe in?

4 What happens to your diaphragm when you breathe out?

5 What happens to the size of your chest when you inhale?

Apply >>

6 Explain why breathing out is usually a passive process.

Extend >>>

7 How does the process of breathing in humans compare to that of other animals, such as fish and insects?

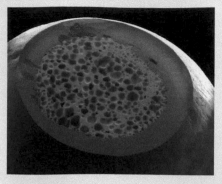

▲ Is this a good model for the structure of alveoli?

▲ A basic model of the alveoli

» Extend: Evaluate models of the respiratory system

You learned earlier in this chapter that the inside of bubbly chocolate looks like alveoli found in your lungs. This is a very simple model of the alveoli. The photo on the left shows a few inflated balloons in a red mesh bag. This is a slightly more complicated model of a cluster of alveoli.

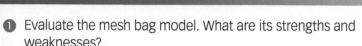

Tasks

① Evaluate the mesh bag model. What are its strengths and weaknesses?

The balloons represent the alveoli. They are the correct shape. The red mesh bag represents the blood capillaries that surround the alveoli. They are the correct colour and approximately the correct thickness when compared to the size of the balloons. In these regards, it is a good model.

However, the balloons do not let air pass through them. If they did, they would immediately deflate and not be balloons! An obvious difference here then is that the lining of the alveoli lets gases through and the latex of the balloons does not. You can trace a path back from every alveolus along bronchioles, a bronchus, the trachea to the mouth. So alveoli in your lungs are connected. In this model the balloons are not connected. Gases cannot pass from one to another.

Bell jar model

The bell jar is quite a common model of the chest and lungs. Many schools have this version. Instructions are available on the internet to make a homemade version of the bell jar model with drinking straws, sticky tape, two balloons and a large drinking bottle. When you blow into the straw at the top of the model the balloons blow up like your lungs.

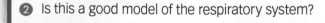

Tasks

② Is this a good model of the respiratory system?

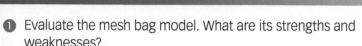

▲ A basic model of the respiratory system

The two balloons represent the lungs. The sections of straw represent the trachea and two bronchi. The shape and position of these are accurate in this model. The bottle itself represents the chest cavity. This is approximately the correct size and shape. Crucially though, you have just learned that the diaphragm moves upwards and intercostal muscles move your ribs upwards and outwards when you breathe in. This is not seen in this model.

Improved bell jar model

The bell jar model can be improved by removing the bottom and adding in a rubber sheet with a handle on.

▲ An improved model of the respiratory system

The rubber sheet represents the diaphragm. When you pull down on it, the balloons in this model expand. This is a good model for inhalation. In real life when your diaphragm contracts it moves downwards and you breathe in.

Tasks

3 Evaluate this model. How has it improved?
4 How could we improve this model even further?
5 Why would filling the inside of the jar (not the balloons) with water improve it?

The mass of the human body is about 65% water. The cytoplasm of each of our cells is made mainly from water. So filling the inside of the jar (but not the balloons) does improve the model. It makes the movement of the rubber sheet and expansion of the balloons more realistic.

Enquiry:
Investigate a claim linking height to lung volume

Go to the Wikipedia™ page for 'lung volume' and you see it says that the average total lung capacity of an adult human male is around 6 litres. It also shows this table:

Larger volumes	Smaller volumes
Taller people	Shorter people
People who live at higher altitudes	People who live at lower altitudes
Non-obese	Obese

But how do we know this is true? How do you measure the volume of your lungs? And do taller people have larger lungs than shorter people?

Measuring lung volume

Aim

To accurately measure the volume of your lungs.

Equipment

Large bottle (more than 5 litres ideally), smaller bottle (ideally 250 cm³), long length of piping, permanent marker, sink full of water.

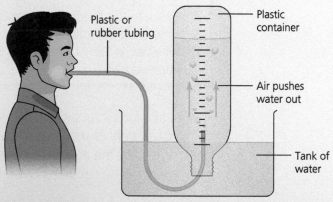

Plastic or rubber tubing

Plastic container

Air pushes water out

Tank of water

How to measure your lung volume ▶

Method

1 Use the smaller bottle to fill the larger bottle of water. Each time you add 250 cm³ to the larger bottle, draw a line at its edge to make a scale. Repeat until you have lines up the complete edge of the large bottle.

2 Put your hand over the top of the large bottle of water. Turn it upside down and put it in the sink of water. Providing you don't lift the bottle from the water it should stay full.

3 Put the tube into the open end of the bottle under water.

4 Disinfect the end of the tubing. Then take a deep breath and blow into it. As you do you will expel water from the bottle. The total amount of water expelled is the volume of your lungs.

5 Repeat until you get three similar results. Ignore odd results and calculate a mean value.

If you are after a more accurate (and drier) method of measuring the volume of your lungs you can use a lung capacity test. These are larger pieces of equipment and look a little like the breathalysers the police use to test for drunk driving. You blow into one end and the machine calculates your lung volume.

▲ The accurate measurement of lung volume in hospital

But do taller people have larger lungs?

In order to answer this question you will have to complete a scientific survey. This will involve asking some of your friends and family for help. You will need to measure the people's height and then follow the method above to measure their lung volume.

Results

Name of person	Height (cm)	Lung volume (cm³)

Try to select people with a range of heights. The more people you study the more confident you can be in your results.

Presenting results

Draw your results as a scatter plot on graph paper. Remember to put height on the *x*-axis and lung volume on the *y*-axis. Also remember to label your axes and include units.

Now try to draw a line of best fit. This can be straight or curved. You don't need to go through every point but are trying to show the general trend of the points. This line shows you the relationship between your two variables. We call this relationship a '**correlation**'.

If your individual results are all close to your line of best fit, then you can be more sure there is a correlation (relationship) between height and lung volume. Are they? If they are not all close to the line then, either there is no relationship or you haven't collected enough data.

If the line of best fit moves upwards from left to right, then we say there is a positive correlation between height and lung volume. As height increases so does lung volume. If the line goes downwards as you go from left to right, then we say there is a negative correlation. As height increases lung capacity decreases.

?

❶ What is a correlation?
❷ In what way does the line of best fit move in a positive correlation?
❸ What are the units of volume?
❹ Was Wikipedia correct? Did you find a relationship between height and lung volume?

Your knowledge objectives

In this chapter you will learn:
- what a balanced diet is and why it is important
- that the function of the digestive system is to break large food molecules into smaller ones
- that iron is important for red blood cells and calcium is needed for strong teeth and bones
- that vitamins and minerals are needed in small amounts to keep the body healthy

See page 171 for the full learning objectives.

▲ Is this a balanced diet?

▲ This person is exercising vigorously

» Transition: Healthy living

We all know that it's important we lead a healthy life. This means we will usually live longer and be happier whilst we do. Many doctors now believe that our mental and physical health are closely linked. But what exactly do we need to do to lead a healthy lifestyle?

Diet

Your diet is a very important part of your lifestyle. There is a famous saying "you are what you eat". What does this mean? The answer is that it means if you eat healthily you will be healthy. Eating a healthy diet means having the correct amounts of certain food types each day. The NHS advise that we should:

- eat five portions of fruit or vegetables every day
- base our meals on starchy foods such as potatoes, bread, rice and pasta
- have some dairy products but not too much
- eat some beans, fish, eggs, meat or other proteins
- reduce the amount of fats
- drink plenty of fluids.

We say that people who eat healthily have a balanced diet. You will learn more about this on the following pages.

Exercise

Regular and vigorous exercise is essential for a healthy lifestyle. Eating healthily and exercising regularly stops us storing extra sugar as fat, and so keeps our weight at an appropriate level. The NHS says that exercise reduces our risk of a major illness like heart disease or cancer by up to a half. It can lower our risk of an early death by up to one-third.

Young people (aged between five and eighteen) should do at least an hour of physical activity per day. This can range from easier activities like cycling and playing in the playground, to harder activities like running and tennis. This exercise also helps you build your muscles and bones. Are you doing enough?

Drugs

Drugs are substances that are taken to cause a change in a person's body or mind. There are many good drugs, which we tend to call medicines. Unfortunately, there are also many illegal drugs that have negative effects on a person's health. Some drugs like cocaine and ecstasy speed up your heart rate and make you feel energetic. These are called **stimulants**. Others like cannabis slow down your heart rate and make you feel relaxed. These are called **depressants**. They don't make you feel depressed, they depress your reaction times. Some drugs like paracetamol are **painkillers**. The final group of drugs are called **hallucinogens**. These affect the way you sense the world around you and can cause worrying hallucinations. These are visions or sounds that are not real.

Confusingly, some drugs like alcohol and tobacco are legal but can cause harm. In small quantities alcohol helps people to relax. Drinking too much alcohol can lead to a lack of self-control. Longer-term effects of drinking too much alcohol include damage to the liver and brain. Many alcoholic drinks can also be fattening.

We have known for a long time that smoking tobacco is very bad for your health. Nicotine is the addictive chemical in cigarettes. Another chemical in cigarettes called tar causes lung cancer. Smoking also causes heart attacks and strokes. If a pregnant women smokes, it can cause miscarriage, premature birth and underdeveloped babies.

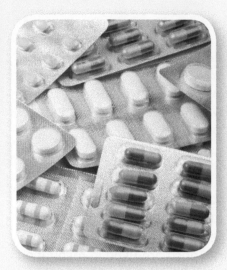
▲ Are all drugs bad? What about these medicines?

▲ Smoking harms you and those around you

Worked example

Who might eat foods that are high in protein, and why?

Protein in our diet is used for growth and repair. Children who are growing or people who are recovering from injury need a high protein diet. Athletes also have high protein diets, to help them recover from training faster.

Know >

1 How many portions of fruit and vegetables should we eat per day?
2 What chemical in cigarettes causes cancer?
3 What chemical in cigarettes is addictive?
4 How much exercise should a young person do each day?

Apply >>

5 What effects does vigorous exercise have on the body?
6 Why can't you drink alcohol and drive?

» Core: Food groups and a balanced diet

There are six key food groups. These are: carbohydrates, proteins, lipids, fibre, vitamins and minerals. Water is another key component of our diet but is not considered a food group.

Carbohydrates

Key word

Carbohydrates are the body's main source of energy. There are two types: simple (sugars) and complex (starch).

Carbohydrates are our main source of energy. Often footballers or other athletes will have a meal full of carbohydrate several hours before they compete. Complex carbohydrates like starch are broken down into simple sugars. These are used in respiration to provide us with energy. You will learn about respiration in Chapter 17. The energy from these sugars allows athletes to compete at the optimum.

Good sources of carbohydrates are cereals, bread, pasta, rice and potatoes. The NHS advise us that starchy foods should make up just over one-third of everything we eat. Our meals should be based upon this type of food. If we eat too many carbohydrates and do not exercise enough, the excess sugar from their breakdown can be stored in our bodies as fat. Over time, excess sugar can lead to obesity, which can then lead to other medical issues such as heart problems.

Proteins

Key word

Proteins are nutrients your body uses to build new tissue for growth and repair. Sources are meat, fish, eggs, dairy products, beans, nuts and seeds.

Proteins are used by your body to make new cells for growth and repair. Good sources of protein are fish, meat, eggs, beans, nuts, seeds and dairy products. Some vegetarians eat beans, nuts and seeds to replace the protein they miss from not eating meat.

▲ Good sources of carbohydrates

▲ Good sources of protein

Lipids

Key word

Lipids (fats and oils) are a source of energy and are found in butter, milk, eggs and nuts.

Lipids are fats and oils. Oils are liquid at room temperature and fats are solid. Lipids are a good source of energy. They also act as a store of energy in the body and insulate it against the cold. Good sources of lipids are butter, oils and nuts.

Fats are saturated or unsaturated. Saturated fats are found in butter, pies, cakes, biscuits, sausages, cheese and cream. Eating too much saturated fat causes a build-up of a special type of fat called cholesterol in your blood. This can lead to heart disease.

Fibre

Key word

Dietary fibre is the parts of plants that cannot be digested, which helps the body eliminate waste.

Dietary fibre is not digested. It passes through your digestive system and is excreted in your faeces. It is very important that we have fibre in our diet, however. It makes our faeces solid. Without fibre, the muscles of our digestive system find it much more difficult to push our food through. Fibre is present in fruit and vegetables.

▲ Good sources of fibre

Vitamins

We need vitamins (and minerals) in much smaller quantities than any of the previous food groups. This doesn't mean they are less important. We need them in small quantities to keep our bodies healthy. Vitamins are named after letters in the alphabet and each has a different reason why we need it.

Vitamin A is needed by your immune system and to see in low light. Vitamin B helps your nervous system. Vitamin C protects your cells. Vitamin D is needed for healthy bones and teeth. Vitamin E helps your skin and eyes. Vitamin K is needed to help prevent blood clots. Vitamins are found in dairy foods, fruit and vegetables.

People with a diet without enough vitamins and minerals can develop deficiency diseases. An example of this is scurvy, which is caused by not having enough vitamin C. This disease was common hundreds of years ago in sailors who were at sea for months without fresh fruit and vegetables. Scurvy causes bleeding gums, loss of teeth, and joint pain.

Minerals

As with vitamins, we need minerals in small quantities and for specific reasons. Calcium helps build strong bones and teeth. Iodine helps you make hormones. Iron helps you make red blood cells which carry oxygen from your lungs to the rest of your body. Minerals are found in salt, milk (for calcium), sea fish (for iodine) and liver (for iron).

Balanced diet

A balanced diet provides us with the correct amounts of each different food group to remain healthy. This will be different for different people. Active younger people who are still growing will need more carbohydrate for energy and protein for growth and repair. They are also likely to need more calcium to help their bones and teeth grow healthily.

Some people try to generalise the amounts of food needed in a balanced diet. They show it in the form of a food plate, as shown below. You can see there is a large section of fruit and vegetables, and another of bread, rice, pasta and other starchy foods. The size of the sections for meat, fish, eggs, milk and dairy foods is much smaller. This approximates a balanced diet.

▲ A food plate showing a balanced diet

We often hear that we should eat five pieces of fruit or vegetables each day. This is to provide us with natural plant sugars, and key vitamins and minerals. Other tips to keep us healthy are:

- cut down on saturated fat and sugar
- reduce the amount of salt you eat
- drink plenty of water
- don't skip breakfast.

A balanced diet provides us with enough energy for an active lifestyle. Without this people become underweight. Overtime this can eventually lead to starvation.

> **Key facts**
>
> Vitamins and minerals are needed in small amounts to keep the body healthy.
>
> Iron is a mineral important for red blood cells.
>
> Calcium is a mineral needed for strong teeth and bones.

Know >

1 What are good sources of carbohydrates?

2 What are proteins used for?

3 What builds up in the blood vessels of people with a high saturated fat diet?

4 What does Vitamin D help with?

Apply >>

5 Why do athletes eat carbohydrate meals before they exercise?

6 Where do vegetarians get their protein from?

Extend >>>

7 What are deficiency diseases? Write a report showing the effects of some of these medical conditions.

>> Core: The digestive system

The digestive system breaks down large lumps of insoluble food into smaller soluble pieces that can be absorbed into the blood. The key parts of the digestive system are shown below.

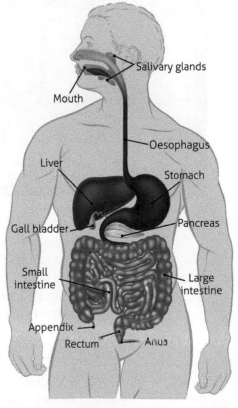

▲ The parts of the human digestive system

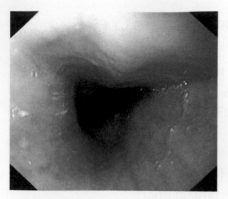

▲ The oesophagus takes food from your mouth to your stomach

Key word

The **stomach** is a sac where food is mixed with acidic juices to start the digestion of protein and kill microorganisms.

The stomach mixes food with acid to ▶ kill pathogens and begins the digestion of protein

Mouth

Here food is broken down mechanically by your teeth as you chew it. The food is mixed with saliva, which is produced by glands in your cheeks. Saliva acts as a lubricant, allowing the food to move down your oesophagus without scratching it. Saliva also contains an enzyme which starts breaking down carbohydrates. Enzymes are biological catalysts which speed up reactions and you will learn more about them later in this chapter.

Oesophagus

Your oesophagus is a thin tube about 30 cm long that connects your mouth and stomach. The lining of the oesophagus can expand to allow you to swallow larger lumps of food.

Stomach

Your **stomach** is a bag about the size of your fist. It has ridges on the insides which allows it to expand when you eat. It contains stomach acid, but this doesn't break down food – it kills any pathogens on your food that might make you ill. So it keeps you safe from illness. In your stomach, food is mixed with another enzyme. This time protein is broken down.

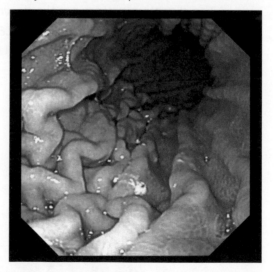

Pancreas

Your pancreas is part of your digestive system but food does not flow through it like many other parts. Your pancreas is a gland that produces enzymes and pumps them into the beginning of the small intestine. It produces enzymes which break down carbohydrates, proteins and fats.

Liver

Like your pancreas, food does not actually flow through your liver. It produces bile which breaks down fats into smaller droplets. Bile is not an enzyme.

Key words

The **small intestine** is the upper part of the intestine where digestion is completed and nutrients are absorbed by the blood.

The **large intestine** is the lower part of the intestine from which water is absorbed and where faeces are formed.

Small intestine

Your **small intestine** is actually over 4 m long. It is called your small intestine because it is narrower than your large intestine. Your small intestine produces three enzymes that break down carbohydrates, proteins and fats. It also absorbs broken down, soluble molecules of food into your blood. Your small intestine has millions of tiny projections called villi, which massively increase its surface area. This means lots of food can be absorbed into the blood.

Large intestine

Your **large intestine** absorbs water from your food. This leaves behind only fibre – the part of food that cannot be digested. Fibre is removed from the body in solid faeces.

Anus

We can think of our digestive systems as a long tube that begins at our mouth and ends at our anus. Your anus is found at the very end of your digestive system. Its function is to control the removal of solid faeces from your body when you go to the toilet.

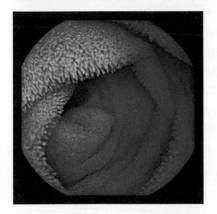

▲ The small intestine absorbs broken down food into the blood

Know ❯

1 Where does digestion start?

2 What two roles does saliva have?

3 What are enzymes?

4 What enzymes are found in your mouth and stomach?

Apply ❯❯

5 What part of the digestive system is working incorrectly if you have diarrhoea?

Extend ❯❯❯

6 How are villi in the small intestine adapted for their function?

7 Design a diet suitable for an athlete and one suitable for a pregnant woman.

❯❯ Core: Digestion

Digestion is the process of breaking down food into smaller pieces. There are two main types of digestion – mechanical and chemical.

Mechanical digestion

Mechanical digestion only happens in your mouth. As you chew, your teeth break down large pieces of food into smaller ones.

The eight teeth at the very front of your mouth are called incisors. These teeth have a sharp edge which means they can bite off lumps of food to then chew. Imagine taking a bite of an apple. It is your incisors that do the biting. The twelve teeth towards the back of your mouth are called molars. These do not have a sharp edge, but a much flatter top surface. As you chew your food you grind it between the two flatter surfaces of your molar teeth.

Chemical digestion by enzymes

We cannot absorb complex carbohydrates, proteins or fats into our blood. They are too large to pass through the lining of the small intestine, so we need to digest them first. You have just learned that mechanical breakdown occurs in the mouth. But the lumps of food that have been chewed and mixed with saliva are still too large to be absorbed. We need enzymes to break them down even further.

Enzymes are biological catalysts that speed up reactions. The enzymes in your digestive system break down larger molecules of food into smaller ones. There are three main types of digestive enzyme:

- carbohydrase enzymes break down complex carbohydrates into sugars

- protease enzymes break down proteins into amino acids

- lipase enzymes break down fats into fatty acids and glycerol.

We are able to absorb the breakdown products of sugars, amino acids, fatty acids and glycerol into our blood.

Carbohydrase enzymes are made in your mouth, so the digestion of carbohydrate begins when saliva containing these enzymes is mixed with your food. Carbohydrase enzymes are also made in your pancreas and small intestine. Protease enzymes are made in your stomach, so the digestion of proteins starts here. They are also made in your pancreas and small intestine. Lipase enzymes are made in your pancreas and small intestine only.

Chemical digestion by bile

Bile is a green-coloured alkaline liquid made in your liver. It is then stored in your gall bladder before being released into the small intestine. Here it neutralises any stomach acid and breaks down fats into smaller droplets. We say that bile **emulsifies** fats. Bile massively increases the surface area of fats. This means that the lipase enzymes have more surface area to act on and so then are able to digest the fats more easily into fatty acids and glycerol.

Chemical digestion by bacteria

It might amaze you to learn that there are more bacteria on your skin and inside your digestive system than you have cells in your body. This means there are millions of billions of bacteria living on and in you right now! But what do these bacteria do?

Some of the bacteria that live on us and in us are pathogens and can cause disease if their numbers increase too high. Other bacteria help us. The bacteria in your digestive system, or **gut bacteria**, do two key things:

- break down some foods like fibre that are impossible to break down with enzymes

- they live in areas where harmful bacteria could grow and cause illness, so there is less space for the harmful bacteria.

> **Key word**
>
> **Gut bacteria** are microorganisms that naturally live in the intestine and help food break down.

▲ Bacteria (coloured green) on the inside of the small intestine

▲ Some foods contain bacteria to 'top up' those found in your digestive system

---

Know >

1 What are the teeth in the front of your mouth called?

2 What do your molar teeth do?

3 What do carbohydrase, lipase and protease enzymes do?

4 What are carbohydrates, proteins and fats made from?

Apply >>

5 Why might you lose weight if you couldn't make bile?

Extend >>>

6 How is bile like washing-up liquid?

7 What would happen if all the enzymes in your body were removed?

» Extend: Critique claims for a food product or diet

We find nutritional information on the packaging of many foods we buy. The companies that make the food we buy have to put this information in view. This is so we can make informed decisions about the food before we put it in our trollies. But what exactly does this information show us?

Nutrition				
Typical values	100g contains	Each slice (typically 44g) contains	% RI*	RI* for an average adult
Energy	985kJ 235kcal	435kJ 105kcal	5%	8400kJ 2000kcal
Fat	1.5g	0.7g	1%	70g
of which saturates	0.3g	0.1g	1%	20g
Carbohydrate	45.5g	20.0g		
of which sugars	3.8g	1.7g	2%	90g
Fibre	2.8g	1.2g		
Protein	7.7g	3.4g		
Salt	1.0g	0.4g	7%	6g

This pack contains 16 servings
*Reference intake of an average adult (8400kJ/2000kcal)

▲ The nutritional food label on the side of loaf of white bread

Nutritional facts labels

These are a type of more traditional label. They are usually black and white and have lots of small numbers on them.

The labels usually start by showing the energy they contain. This is given in both kilojoules (kJ) and kilocalories (kcal). Both are measures of the energy that is inside the food. Many people tend to measure energy in their food in calories, but kilojoules are a more scientific unit to use.

Below this the food label has the amount of fat the food contains. This is measured in grams. The labels then give the amount of total fat that is saturated. You will remember that saturated fats cause a build-up of fat in blood vessels which can lead to heart attacks and strokes. The label then shows the amount of carbohydrates. This is also measured in grams. Again, the amount of carbohydrate that is sugar is given. Finally, the label tells us the amount of fibre, protein and salt.

Towards the right-hand side of the label is a column called '% RI'. This stands for percentage of a person's daily recommended intake (RI). This is the percentage that one serving would give as a part of the person's diet. So the label above says that a serving of bread gives us 5% of the daily recommended intake of energy and 1% of the fat.

Traffic lighting panels

Each 1/2 pack serving contains

MED	LOW	MED	HIGH	MED
Calories	Sugar	Fat	Sat Fat	Salt
353	**0.9g**	**20.3g**	**10.8g**	**1.1g**
18%	1%	29%	54%	18%

of your guidline daily amount

▲ The traffic lighting panels

Some people find it difficult to understand the nutritional information panels, so food manufacturers have tried to make this easier by introducing a traffic light panel. It is often on the front of the food packaging to make it even easier. This has a summary of some key information from the nutritional fact panel. Below this the same percentages of your daily intake are given, but they have been colour coded. Green on the label means the food is low in this substance. The label on the previous page is green for sugar and so has a low, healthy level of this. An orange colour means a food has higher levels and a red means it is high in this substance. The label we've been looking at shows that the food contains medium levels of salt and fat and high levels of saturated fat.

Simply put, the more green you see, the more healthy the food is and the more red, the less healthy.

How healthy are these labels?

The following labels have come from four foods. What do they tell you about them? Can you work out what type of food they are from?

Nutrition

Typical values	100g contains	Each pack contains	% RI*	RI* for an average adult
Energy	1083kJ 258kcal	1865kJ 445kcal	22%	8400kJ 2000kcal
Fat	11.9g	20.4g	29%	70g
of which saturates	4.4g	7.6g	38%	20g
Carbohydrate	22.5g	38.8g		
of which sugars	1.8g	3.1g	3%	90g
Fibre	2.3g	3.9g		
Protein	14.3g	24.7g		
Salt	1.1g	1.8g	30%	6g

Pack contains 1 serving

*Reference intake of an average adult (8400kJ / 2000kcal)

NUTRITIONAL INFORMATION

	per 100g	per 1/2 Pizza
Energy	927kJ 220kcal	1400kJ 333kcal
Fat	7.2g	10.9g
of which Saturates	3.2g	4.8g
Carbohydrates	28.1g	42.4g
of which Sugars	3.6g	5.4g
Fibre	1.6g	2.4g
Protein	10.0g	15.1g
Salt	1.15g	1.74g

This pack contains 2 servings

Energy 414kJ 98kcal	Fat 0.4g	Saturates 0.1g	Sugars 1.1g	Salt 0.81g
5%	1%	1%	1%	14%

of your reference intake
Typical values per 100g: Energy 753kJ/178kcal

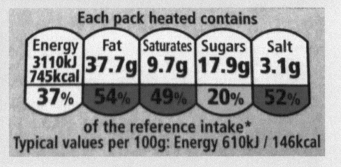

Each pack heated contains

Energy 3110kJ 745kcal	Fat 37.7g	Saturates 9.7g	Sugars 17.9g	Salt 3.1g
37%	54%	49%	20%	52%

of the reference intake*
Typical values per 100g: Energy 610kJ / 146kcal

 ▲ The food labels for four packets of food

Tasks

1. What are the two units for energy?
2. What is fat in foods measured in?
3. What does % RI stand for?

Enquiry:
Evaluate models of the digestive system

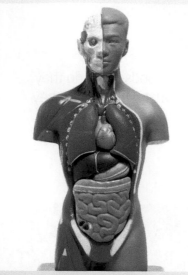

▲ A traditional school model for the digestive system

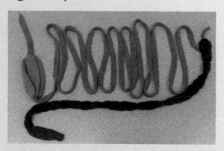

▲ This is a knitted model of the digestive system

▲ Spaghetti in a pair of tights is also a model for the digestive system

Model of human torso

You may have seen a model like this of a human body. What organs can you see? The lungs are visible either side of the heart. Below that the dark-brown section is the liver and the paler feather-shaped part is the pancreas. Below that the grey section is the large intestine and the more central pink part is the small intestine. But is this a good model of the digestive system?

You may have spotted that organs of other systems can be seen. The heart is part of the circulatory system. Models like this tend to be very realistic. The sizes, shapes and basic colours of the organs are correct. The model allows you to remove organs and see what is behind. However, the organs themselves are solid and rigid. They do not move. Your small intestine is a long thin tube about 4.5 m long. It is difficult to tell this from the model.

Knitted digestive system

The knitted digestive system is another model. This has many of the same advantages as described above. But it has an advantage. This version is again life-sized but here the small intestine is actually a thin tube 4.5 m long. In this regard, this model is more realistic.

However, neither of these two models show you anything about how the organs work.

Spaghetti in tights

In this model the pair of tights is the lining of the small intestine – the spaghetti inside is the food inside the intestine. You can see in this model that bits of spaghetti pasta represent the large pieces of food we eat. These are broken down into much smaller pieces, which in this model are the sauce. The sauce can leave the tights, just as food can leave the small intestine. So far, this seems a good model. However, broken down food leaves the small intestine and enters the blood. Where is the blood in this model?

Visking tubing

▲ Visking tubing is a very scientific model for the digestive system

Perhaps the most effective model for the digestive system is Visking tubing. This is a special type of tubing that allows small molecules like the sugar, glucose, to pass through it but stops large molecules like starch passing through. In this regard it is exactly the same as the lining of the small intestine. This won't let starch though but will let glucose pass through after it has been broken down from starch. Look at the diagram above. Where is the blood? Well it must be on the other side of the tubing. So it has to be the water that surrounds it.

❶ What is starch broken down into?
❷ What advantages does each model have?
❸ What disadvantages does each model have?
❹ How could you improve each model?
❺ Can you think of a better model?

Ecosystems

Learning objectives

17 Respiration

In this chapter you will learn...

Knowledge

- that respiration is a chemical reaction that releases energy from glucose
- that most living things use aerobic respiration but switch to anaerobic respiration when oxygen is unavailable
- how fermentation of yeast is used in brewing and bread-making
- the definitions of the terms aerobic respiration and anaerobic respiration

Application

- the word equations for aerobic and anaerobic respiration
- how to explain how specific activities involve aerobic or anaerobic respiration

Extension

- how to suggest how organisms living in different conditions use respiration to get their energy
- how to describe the similarities and differences between aerobic and anaerobic respiration

18 Photosynthesis

In this chapter you will learn...

Knowledge

- that plants and algae obtain their energy from photosynthesis
- that these organisms use the glucose produced as an energy source
- the organs needed for photosynthesis
- that iodine is used to test for starch
- the definitions of the terms fertilisers, photosynthesis, chlorophyll and stomata

Application

- how to describe ways in which plants obtain resources for photosynthesis
- how to explain why other organisms are dependent on photosynthesis
- how to draw a line graph to show how the rate of photosynthesis is affected by changing conditions
- the word equation for photosynthesis

Extension

- how to suggest how particular conditions could affect the growth of plants
- how to suggest why leaves, roots and stems are adapted
- how to explain the movement of carbon dioxide and oxygen through stomata at different times of day

17 Respiration

Your knowledge objectives

In this chapter you will learn:
- that respiration is a chemical reaction that releases energy from glucose
- that most living things use aerobic respiration but switch to anaerobic respiration when oxygen is unavailable
- how fermentation of yeast is used in brewing and bread-making
- the definitions of the terms aerobic respiration and anaerobic respiration

See page 199 for the full learning objectives.

» Transition: Staying alive

What do you need to stay alive? Have you ever wondered this? As an animal, you are now old enough to not rely upon your parents. (If you are under 18 in many countries you are legally dependent upon your parents.) But what exactly are your basic needs? What do you need to stay alive? Our needs are the same as almost all other animals on our planet.

Air

Nothing can live in space. Most of it is a vacuum, which means there is no air (or any other type of atmosphere). Without air, there is no oxygen and without it life cannot survive. It is entirely possible that life exists in liquid or frozen water on other planets, moons or other objects in space. But it cannot live in the vacuum of space.

Water

Water is found everywhere on our planet. It is frozen in the North and South Poles. It shoots out from geysers at extremely high temperatures and pressures. It is widely known that species like polar bears live near the North Pole and penguins near the South Pole. But bacterial life also exists in the superheated geysers. Wherever we have found water, we have also found life.

Water is important for life because many substances can dissolve in it. Oxygen dissolves in liquid water and so allows aquatic life like fish to survive. Glucose also dissolves in it, so when the food you eat is broken down, the glucose dissolves into your blood and is transported around your body.

Water is so precious that many species have evolved adaptations to keep as much of it as possible. Many desert animals burrow into the cool sand during the day to save water. Cacti have spines instead of leaves to reduce water loss. Many animals including ourselves produce more concentrated urine on hotter days to save water. Have you ever noticed that your urine is darker in colour (so more concentrated) when you are dehydrated?

▲ No life can live in the vacuum of space

◀ Life can survive in some very cold and some very hot places on our planet

Food

All living organisms need food. Some organisms like plants and algae can make their own food by photosynthesis. You will learn more about this in the following chapter. All animals get their food from eating plants if they are **herbivores** or other animals if they are **carnivores**. Some animals like bears eat both plants and other animals and are called **omnivores**. Food provides organisms with enough:

- sugars and fats for respiration

- proteins for growth and repair

- vitamins and minerals to stay healthy

- water to not become dehydrated.

Some reptiles like snakes can go long periods without food. They can often wait days or even months between feeds. They are able to slow the reactions in their bodies down to survive for long periods without food. Other animals like the polar bear are able to sleep for months on end. When food stocks are low, such as in the deep winter, many animals including the polar bear hibernate. During this deep sleep their heart rate, temperature and breathing all slow to save energy. They wake when food becomes plentiful again.

▲ After eating this rat, the python may not feed for many weeks

Worked example

Why do animals hibernate?

Animals hibernate during the winter to save energy. There is often not enough food around to keep them alive.

Know >

1 Why do you need air to remain alive?

2 Besides oxygen, what else do you need to stay alive?

3 How do many desert animals reduce their body temperature during the day?

4 Why do we need protein in our diet?

Apply >>

5 What colour is your urine when you have drunk sufficient water?

6 What other animals are omnivores?

» Core: Aerobic respiration

You are made from around 35 million billion cells. Some of these are skin cells, others are nerve, muscle or blood cells. In fact, there are around 200 different types of cell in your body. All of the cells that make up you and me, and in fact every living organism on our entire planet, are respiring now. Respiration is one of the key processes for all living things. Without it individual cells, and then entire organisms, will die.

Aerobic respiration

Although it sounds like it, aerobic does not mean in the presence of air. It specifically means in the presence of oxygen. Aerobic respiration is a series of reactions that release energy stored in glucose. The respiring organism then uses this energy to complete the seven life processes: movement, respiration, sensitivity, nutrition, excretion, reproduction and growth. The word equation for this aerobic respiration is:

$$\text{glucose} + \text{oxygen} \xrightarrow{\text{releases energy}} \text{carbon dioxide} + \text{water}$$

The balanced symbol equation for this is:

$$C_6H_{12}O_6 + 6O_2 \xrightarrow{\text{releases energy}} 6CO_2 + 6H_2O$$

Aerobic respiration is an exothermic reaction because it releases energy.

Respiration happens all the time in all cells of an organism. If respiration stops happening, that cell will die. An entire organism will die if too many of its individual cells die. The five kingdoms of life on Earth are animals, plants, bacteria, fungi and protists. Every cell of every organism in these kingdoms respires all the time.

It is especially important to remember that it is not just animals that respire, but plants also. Plants photosynthesise during the day, but respire during both the day and night.

Reactants and products

The two reactants in respiration are glucose and oxygen. You can remember these by asking yourself why you breathe and why you eat. You breathe to provide your respiring cells with oxygen and you eat to provide these cells with glucose for respiration (and other key nutrients).

Key word

Aerobic respiration is the process of breaking down glucose with oxygen to release energy and produces carbon dioxide and water.

▲ All life in this rainforest, including the trees, respires all the time

▲ Water is produced during respiration and is breathed out. Here it can be seen condensing on a window

The products of respiration are carbon dioxide and water. Most people remember we breathe in oxygen and breathe out carbon dioxide. It is important to remember that we do breathe out some oxygen, just less than we breathed in. The reverse is true for carbon dioxide. How do you remember we breathe out water vapour as well? Imagine yourself breathing onto a window in winter. What happens? The answer is the condensation from your breath forms on the window, showing we breathe out water.

Although it is neither a reactant nor product, energy is released during aerobic respiration. You learned earlier that this is used for the seven life processes. It is worth remembering that in warm-blooded animals like mammals, including us humans, and birds, much of this energy is used to provide body heat. Cold-blooded animals like reptiles, amphibians and fish do not use energy from respiration to heat themselves.

Mitochondria

Aerobic respiration occurs in special cell components called mitochondria. These are found in the cytoplasm. You have learned that the cytoplasm is a liquid in which cellular reactions occur. Well, respiration is one of those reactions that happens in mitochondria in the cytoplasm of cells.

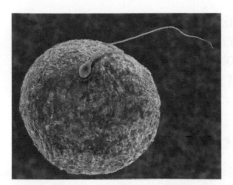

▲ Sperm have many mitochondria to release energy to allow them to reach the egg

Some cells like sperm or muscle cells will have many more mitochondria than others. Why do you think this is? Sperm need to swim to the egg cell and muscle cells need to contract and relax. To be able to do this, they need energy. This comes from respiration. So they have more mitochondria so more repiration occurs.

Know >

1 What is the word equation for respiration?

2 What are the products of aerobic respiration?

3 What does aerobic mean?

Apply >>

4 Why is respiration an exothermic reaction?

5 Where is the energy released during respiration originally stored?

6 Why do you breathe?

Extend >>>

7 What is the balanced symbol equation for aerobic respiration?

8 Why do muscle cells have more mitochondria?

» Core: Anaerobic respiration in humans

Key word

Anaerobic respiration (fermentation) is the process of releasing energy from the breakdown of glucose without oxygen, producing lactic acid (in animals) and ethanol and carbon dioxide (in plants and microorganisms).

At some points in our lives we are unable to provide our cells with all the oxygen they need to complete aerobic respiration. This is likely to be towards the end of long sports events like the cross-country. Here your cells have been respiring to produce energy for long periods of time. Your red blood cells cannot carry enough oxygen from your lungs to your cells. So, as well as you feeling very tired, what is happening?

When we don't have enough oxygen for aerobic respiration our cells respire anaerobically. Again, this does not mean without air, but specifically means without oxygen. The word equation for **anaerobic respiration** is:

$$\text{glucose} \xrightarrow{\text{releases energy (5\%)}} \text{lactic acid}$$

Both aerobic and anaerobic respiration are **exothermic** because they release energy. However, there is a crucial difference here. Aerobic respiration only releases about 5% of the energy when compared with aerobic respiration. At the point that you really need more energy (like the end of your cross-country run), your body is unable to provide it.

▲ This athlete's muscles will need glucose and oxygen for respiration during his race

Reactants and products

As with aerobic respiration, you get your glucose for anaerobic respiration from your diet. You often feel tired when you haven't eaten enough. This is because your cells cannot respire enough to provide you with energy.

The key difference between aerobic and anaerobic respiration is that the lack of oxygen as a reactant in anaerobic respiration means that glucose cannot be fully broken down. You learned on the previous page that the products of aerobic respiration are carbon dioxide and water. Both of these contain oxygen. But in anaerobic respiration, there is not enough oxygen. So this means that the reaction can only be partly completed. Glucose can only be partly broken down into lactic acid. It cannot be fully broken down into carbon dioxide and water. Lactic acid is a toxin and causes cramp when it builds up in your muscles.

Oxygen debt

When you owe a friend a favour you are in their debt. When your body is respiring anaerobically we say that you owe your body an oxygen debt. This is also called **EPOC (Excess Post-Exercise Oxygen Consumption)**.

After you've finished the cross-country race your muscle cells have been respiring anaerobically and you owe them oxygen. Your breathing rate stays high even when you've finished. As well as this you are breathing more deeply than normal to repay the debt. Your pulse rate stays high to pump the oxygen to the cells that need it. But what happens to the oxygen when it gets there? The following reaction happens:

▲ These athletes have produced lots of lactic acid during the race

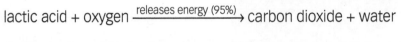

lactic acid + oxygen $\xrightarrow{\text{releases energy (95\%)}}$ carbon dioxide + water

So the lactic acid is broken down, which means your cramp or stitch disappears. The rest of the energy is released so you start to feel more energetic again.

Know >

1 What is the word equation for anaerobic respiration?

2 What does anaerobic mean?

3 What is an oxygen debt?

4 What proportion of energy is released during anaerobic respiration compared with aerobic respiration?

Apply >>

5 Where is the remaining 95% of the energy that is not released during anaerobic respiration?

6 How is this energy finally released?

7 How does your body try to repay your oxygen debt?

8 What would happen if you didn't produce lactic acid during anaerobic respiration?

» Core: Anaerobic respiration in microorganisms

There are three types of microorganisms: bacteria, fungi and viruses. Only bacteria and fungi respire. Viruses do not and so are not considered alive. Yeast is an incredibly important fungus because of the way it respires. We use yeast to bake bread and brew beer. Both of these are hugely important foods economically. Before we were able to sterilise water to make it safe to drink, beer was drunk by many people in place of water. The process of turning it alcoholic actually helped make it safe to drink!

When fungi like yeast respire, they do so anaerobically. This process is not the same as in animals like us, which you have just learned about on the previous page. It is without oxygen so we called it anaerobic, but we also call it **fermentation** to show it is different from anaerobic respiration in animal cells like our own.

The word equation for fermentation (anaerobic respiration in yeast) is:

$$\text{glucose} \xrightarrow{\text{releases energy}} \text{ethanol} + \text{carbon dioxide}$$

The balanced symbol equation for this is:

$$C_6H_{12}O_6 \xrightarrow{\text{releases energy}} 2C_2H_6O + 2CO_2$$

Reactants and products

As with anaerobic respiration in humans on the previous page, the only reactant is glucose. When baking bread, we add glucose in the form of sugar. When brewing beer, the sugar is released from the malted barley when it is soaked. This is the first step in the process of brewing.

Yeast needs the sugar added to bread or released from the barley to live. To do so it must respire, so it breaks down the glucose and releases energy. In doing so it produces ethanol as a waste product. Ethanol is a type of alcohol. It is found in different quantities in all alcohol we drink. Beers are about 4% alcohol, wines are about 12% and spirits like vodka are about 40% ethanol. During fermentation yeast also releases carbon dioxide. This causes bubbles in both our bread and beer.

▲ Bread and beer are popular parts of some adult diets

▲ Yeast is often bought in a dried form for baking

▲ Alcohol is an economically important item

But if yeast produces ethanol when it ferments, why does beer contain ethanol and bread not? Why do we not get drunk when we eat bread? The answer is in its making. Bread is baked at high enough temperatures to kill the yeast before it can make too much ethanol and evaporate off any of this alcohol that has been made. Because we don't heat beer at high temperatures, we don't kill the yeast so quickly and don't stop it producing ethanol.

Importance of bread

Throughout the whole of our recorded human history, bread has been a key staple food. The proportions of the ingredients and indeed the types of flour that have been used have varied significantly, but the process is largely similar. Imagine your life without bread. No toast for breakfast and no sandwiches for lunch. Even the word companion in Latin translates to "with bread" (so a companion is a friend who has bread)!

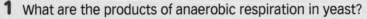

Know >

1 What are the products of anaerobic respiration in yeast?

2 What type of organism is yeast?

3 What foods do we need yeast to make?

4 What does yeast need for fermentation?

Apply >>

5 How does the percentage of ethanol in different types of alcohol differ?

6 Why is bread not alcoholic?

Extend >>>

7 What is the balanced symbol equation for fermentation?

> **Key fact**
>
> Yeast fermentation is used in brewing and breadmaking.

» Extend: Respiration in different organisms

We can investigate respiration in invertebrates experimentally. This experiment involves using small invertebrates like grasshoppers and crickets. It is absolutely essential that these are treated in the very best possible way. No harm must come to them during the experiment.

Aim

To investigate the rates of respiration in different invertebrates.

Method

1 Gently place a small invertebrate like a grasshopper into a boiling tube.

2 Carefully push a small section of cotton wool into the middle of the tube to hold the invertebrate in its end.

3 Place a small mass of soda lime in the tube after the cotton wool.

4 Attach a bung to the boiling tube with a capillary tube with a small drop of water within it.

5 Record the time taken for the water droplet to move 2 cm along the tube.

6 Repeat the experiment using other grasshoppers until you have three similar results.

7 Repeat the whole experiment using another species invertebrate like a cricket.

8 Calculate mean times for the droplet to move and compare results.

Results

The invertebrate with the quickest time for the bubble to move 2 cm is respiring at the fastest rate.

When the invertebrate respires, it takes in oxygen and releases carbon dioxide. The carbon dioxide is absorbed by the soda lime and so removed from the boiling tube. This causes a partial vacuum and so the drop of water in the capillary tube moves towards the invertebrate.

It is essential you stop the experiment by removing the bung before all oxygen in the boiling tube is used up by the respiring invertebrate. Otherwise it will die.

How animals are adapted to extreme conditions: the desert

We all know that it is often very hot in the desert during the day and that it doesn't rain much. In fact, the hottest place on Earth is the desert in Death Valley in the USA where it has reached nearly 58 °C. The Atacama Desert in South America is also the driest place

Experiment 1
Cotton wool · Screw clip · Soda lime · Rubber tubing · Grasshopper · Capillary tube · Water drop

Experiment 2
Cricket

▲ The equipment used to investigate the rate of respiration in two different invertebrates

▲ Camels are adapted to live in the desert

on Earth. Here it hasn't rained since we began keeping records! So how are the animals that live in places like this adapted to survive these extreme conditions?

Animals that do not regulate their body temperature like lizards and other reptiles have an advantage in hot places. They do not need to use the energy released from respiration to heat themselves. They can wait until the Sun's light energy has warmed them and their surroundings before doing their daily activities like hunting. They therefore have more energy from respiration to spend on these activities.

Warm-blooded animals like mammals can struggle to keep cool in the desert. They can spend much of their energy released from glucose during respiration doing so. They often have other adaptations to save energy. Desert foxes have big ears. This increases their surface area to volume ratio to lose heat. The body temperature of camels can increase to 40 °C during the day. This temperature would kill many other organisms.

How animals are adapted to extreme conditions: the poles

The coldest temperature ever recorded on Earth is –89 °C, which occurred in Antarctica in 1983. With the rate of temperature increase because of the greenhouse effect leading to global warming, will this be the lowest ever temperature we record? Although both poles have plenty of frozen water, the availability of fresh (not salty) liquid water is low. There is little rainfall. So how are the animals that live in places like this adapted to survive these extreme conditions?

Many of the animals found in our polar regions are mammals or birds. These are warm-blooded. Cold-blooded organisms like reptiles rely upon their environment to keep them warm and so would find life in the cold poles very difficult.

▲ Polar bears are adapted to live in the Arctic

The polar bear is a mammal. It must use some energy released from glucose in respiration for heat. It also has a thick layer of relatively quick drying fur to keep it warm. Below this is a thick layer of insulating fat. Polar bears have tiny ears. This reduces their surface area to volume ratio to retain heat.

Tasks

1. Where is the hottest place on Earth?
2. What is unusual about the body temperature of camels?
3. Why do many animals in the Arctic have small ears?
4. Why was soda lime used in Experiment 1?
5. What would have happened if you had left Experiment 1 running overnight?
6. What are potential ethical issues with Experiment 1?
7. What size would you expect animals in the Arctic to be? Explain your answer.
8. How do you calculate a mean value?
9. Why do we repeat experiments until we have three similar results?

Enquiry:
Using yeast to explore fermentation

You learned earlier in this chapter that yeast is a fungus that respires anaerobically to produce an alcohol called ethanol. Here again are the equations for this fermentation reaction:

$$glucose \xrightarrow{\text{releases energy}} ethanol + carbon\ dioxide$$

The balanced symbol equation for this is:

$$C_6H_{12}O_6 \xrightarrow{\text{releases energy}} 2C_2H_6O + 2CO_2$$

This experiment measures the volume of carbon dioxide produced at different temperatures. More carbon dioxide will be produced at the optimum temperature for fermentation.

Measuring the carbon dioxide produced

Aim
To determine the optimum temperature for fermentation.

▲ The carbon dioxide released from this respiring yeast can be seen as froth

Equipment
Boiling tubes, yeast solution, sugar solution, balloons, water baths set to 20, 30, 40 and 50 °C.

Method
1 Add 5 cm³ of yeast solution to a boiling tube.

2 Next add 10 cm³ of sugar solution and place a balloon over the top of the boiling tube.

3 Place in a water bath at 20 °C for 20 minutes.

4 Repeat the experiment for water baths at 30, 40 and 50 °C.

5 After 20 minutes compare the size of balloon at each temperature.

Results

The most inflated balloon has the most carbon dioxide in it. This means the yeast has respired the most.

Conclusions

All other variables were kept constant apart from temperatures. So the temperature at which the most carbon dioxide was produced is the optimum one for anaerobic respiration of yeast.

Of the temperatures given above it is most likely to be 30 or 40 °C.

Note

The gas released by the respiring yeast can be shown to be carbon dioxide by bubbling it through limewater. This turns cloudy in the presence of carbon dioxide.

Another way of comparing the rate of fermentation at different temperatures is to measure the height of froth formed.

▲ Carbon dioxide from fermentation turns limewater cloudy

❓

1. What is the word equation for fermentation?
2. What is ethanol a type of?
3. What gas is produced during anaerobic respiration of yeast?
4. Why do microorganisms ferment glucose?
5. Why do organisms like yeast have an optimum temperature?
6. How can you prove that yeast releases carbon dioxide when it respires?

(18) Photosynthesis

» Transition: What do plants need to grow?

The ancient Greeks thought plants grew by taking nutrients from the soil. They did not test this idea scientifically, or they would have found it to be incorrect. Thousands of years later, a Dutch scientist called Van Helmont (1580 to 1644) did test it. He planted a tree in a large pot of soil. He weighed both the tree and soil before starting. He watered the plant for five years. After this time, he weighed the soil again and found it had reduced by a tiny amount. He weighed the tree and found it had increased by a large amount. This was the first in several key experiments that proved that plants are able to make their own 'food' by photosynthesis and do not grow by simply taking nutrients from the soil. So what exactly do plants need to grow?

Air

Nothing living can survive in space. So just like us, plants need air. They need the carbon dioxide in air to complete photosynthesis. You will learn more about this later in this chapter. They also need oxygen in air for respiration. This was described in the previous chapter. These gases move in and out of plants through their leaves.

Light

No plants can grow completely in the dark, and so we don't find any plants deep in the ocean or far beyond the entrances to caves. All plants need some light. They often have leaves that catch this light for photosynthesis. They often have trunks, branches and stems that hold leaves in the correct position to maximise the amount of light that shines on them.

The amount of light that is needed depends upon the plant. Some like ivy are happy to grow in darker places, but others like cacti have evolved to grow in the bright light of deserts during the day.

▲ Cacti have evolved to grow in deserts with bright light and little water

▲ Mangrove trees can grow in water that would kill many other plants

Water

All plants need water. Some like mangrove tress can survive in high levels of water. Many other plants would be killed if their soil was this wet. Others like cacti have evolved to live in drier places like the desert. Water is absorbed through plant roots and transported to the leaves for photosynthesis.

Nutrients from the soil

You need very small amounts of vitamins and minerals from your diet to keep you healthy. So do plants. They need minerals, which they absorb through their roots.

Space

Just like animals, plants need space. If they grow too close to other plants then they compete for the light, water and nutrients from the soil. This often means they cannot grow as quickly or healthily.

▲ Plants can only grow around the entrances to caves because they need light

Worked example

Some farmers have started to plant vegetables very close together on purpose. Can you work out why they do this?

So that they compete with each other. This means that they cannot grow so large. They produce smaller plants called micro vegetables, which are very fashionable in some expensive restaurants.

Know >

1 What process do plants need light for?

2 What happens to plants that do not get sufficient light?

3 How is water absorbed into plants?

4 Why can't plants live in caves?

Apply >>

5 Why do plants need air to grow?

6 Why can only some plants grow in deserts?

7 Why do plants need minerals?

Key words

Photosynthesis is the process that uses the Sun's energy to convert carbon dioxide and water into glucose and oxygen.

Chlorophyll is the green pigment in plants and algae that absorbs light energy for photosynthesis.

Stomata are pores in the bottom of a leaf which open and close to let gases in and out.

▲ Photosynthesis happens in all green parts of plants, so both the leaves and stem

» Core: Photosynthesis

Photosynthesis means 'making from light'. Plants and other photosynthesising organisms like algae possess an amazing ability. They can use energy from light to convert carbon dioxide and water into glucose and oxygen. The word equation for this is:

carbon dioxide + water $\xrightarrow[\text{chlorophyll}]{\text{light energy}}$ glucose + oxygen

The balanced symbol equation for this is:

$$6CO_2 + 6H_2O \xrightarrow[\text{chlorophyll}]{\text{light energy}} C_6H_{12}O_6 + 6O_2$$

Photosynthesis is an **endothermic** reaction because it absorbs energy.

At first some people are not amazed by this ability. But more than 99% of all life on Earth directly depends on the ability of plants and algae to photosynthesise. You, me and all other animals alive depend upon it too. Simply put, without photosynthesising plants almost all life on Earth would die.

Light energy is written above the arrows for photosynthesis. This reminds us that light is essential for photosynthesis. In fact, the reaction cannot happen in the dark – plants are only able to photosynthesise during the day, unless they are lit by artificial light. Below the arrow is written **chlorophyll**, a green pigment. Photosynthesis occurs within green parts of plants. You will learn more about this later in this chapter.

Reactants and products

The two reactants in photosynthesis are carbon dioxide and water. Plants take in carbon dioxide through tiny pores in their leaves called **stomata**. They take in water through specially adapted cells in their roots called root hair cells.

Plants use energy from light to break apart molecules of carbon dioxide and water and make molecules of glucose and oxygen. Plants have evolved to make glucose. Oxygen is the second product of photosynthesis. For plants, oxygen is a simple by-product. They do not try to make it. Thank goodness that they do, because this is essential for all animal life on Earth.

Factors that affect photosynthesis

Anything that can slow the rate of a chemical reaction like photosynthesis is called a limiting factor. Plants require carbon dioxide and water for photosynthesis. So if a plant does not have enough of either of these chemicals then the rate of photosynthesis will reduce. They have become a limiting factor.

Light also affects the rate of photosynthesis. Aquatic plants in fish tanks give off more bubbles of oxygen when their lights are closer to them. This proves that more photosynthesis occurs with more light. Scientifically speaking we say more photosynthesis occurs at higher light intensities.

▲ More bubbles of oxygen are given off when this student moves the light closer to the pondweed

Uses of glucose

The glucose that is made during photosynthesis is often used by the plant as an energy source during respiration. You learned about this in the previous chapter. Glucose is also used to build new cellulose. This is a polymer which surrounds all individual plant cells to provide the plant with an upright structure. Glucose can also be stored as starch to be used later. Potatoes and rice are stores of plant starch which we grow as crops to eat.

Producers and food chains

▲ The energy to feed this lioness originally came from the Sun

At the bottom of almost all food chains is a producer. This is a plant or alga that can complete photosynthesis. This organism makes glucose during photosynthesis. Producers are eaten by primary consumers. These are animals which eat plants. They are called herbivores. They get their energy to live from the plant. The primary consumers are in turn eaten by secondary consumers and so on. These are also animals but they are carnivores because they eat other animals. Therefore all energy in a food chain comes from the photosynthesis completed by the producer. This is how plants and algae are essential for almost all life on Earth.

Know >

1 What are the reactants in photosynthesis?

2 What are the products of photosynthesis?

3 Within a cell, where is chlorophyll found?

4 What else is essential for photosynthesis?

5 What are the three uses of glucose?

6 What is a limiting factor?

Apply >>

7 Why are plants essential for most life on Earth?

8 Why do some farmers use gas burners in their greenhouses?

Extend >>>

9 Why do some fish tanks with plants in have carbon dioxide bubbled into them?

10 What is the balanced symbol equation for photosynthesis?

Enquiry >>>>

11 How can we prove light intensity determines the rate of photosynthesis?

» Core: Leaf structure

You learned on the previous page that photosynthesis in plants and algae provides the energy in almost all of the food chains on Earth. Photosynthesis happens in all green parts of plants. These cells are green because they have chloroplasts inside their cytoplasm. The chloroplasts contain lots of chlorophyll. Many plant stems, especially of smaller plants and cacti, are green and so contain chloroplasts in which photosynthesis occurs. Almost all plant roots are white. They are found under the ground where no light can shine. There is therefore no point in the root cells of most plants having green chloroplasts.

The vast majority of photosynthesis does occur in leaves and so we can think of them as the source of the majority of food on our planet.

Size and shape of leaves

Leaves come in a large variety of sizes and shapes. Plants have evolved these differences to suit the conditions in which they live. You will learn about evolution in Chapter 19. Many leaves like those of the horse chestnut tree are large and thin. Their leaves have a large surface area. This means that lots of the Sun's light energy falls on their leaves meaning they can maximise their photosynthesis. The leaves of many of our trees are so good at catching the light that it is often quite shaded when you stand under them.

Plants that live in hot conditions like cacti in the deserts have spines for leaves. The sunlight is so strong in the deserts that the green cells in the plant stem can complete photosynthesis and its leaves have slowly evolved to protective spines. Other plants called succulents have evolved thicker leaves in which they store much of their water.

The flat surface of a leaf is called a blade. Leaves often have a strong section running down their middle called a midrib. This provides support, holding the leaf in sunlight. Coming from the midrib are often many veins. These move water to and from the leaves.

A leaf in cross-section

The diagram on the next page shows a leaf in cross-section. This means that we have cut through it and are looking inside.

At the top of the leaf is a waxy layer called a cuticle. Plants need water for photosynthesis. For many plants, water is a very precious resource which cannot be wasted. The waxy cuticle on the top stops too much water being lost from the plant.

At the top of leaves are palisade cells. They contain a high number of chloroplasts to maximise the amount of photosynthesis.

▲ The diversity of sizes and shapes of leaves

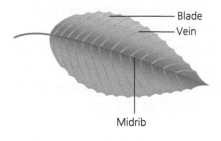

Blade
Vein
Midrib

▲ The structure of a leaf

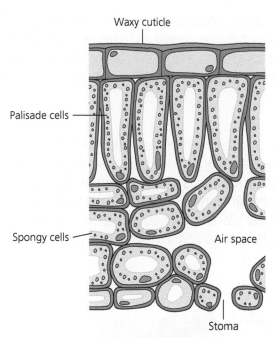

Waxy cuticle

Palisade cells

Spongy cells

Air space

Stoma

▲ The internal structure of a leaf

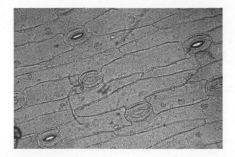

◄ A magnified image of stomata with guard cells surrounding them

Towards the bottom of a leaf cross-section are a layer of spongy cells. These have fewer chloroplasts because less light shines on them. They do have air spaces between them to allow gases in the air to diffuse in and out more easily.

At the very bottom of leaves are special holes called stomata. Around each are two special cells called guard cells. They can swell up to open the stomata or shrink to close it. When leaves have open stomata more gases in air can move by diffusion in to and out from the spongy cells.

Carbon dioxide diffuses into plant leaves through stomata and then into their cells for photosynthesis. This process produces oxygen which diffuses out the same way. But plant cells also respire – oxygen diffuses in and carbon dioxide diffuses out from the cells too. You will learn more about the movement of gases into and out from stomata in the Extend section of this chapter.

You read above that most leaves are flat to increase their surface area to absorb more energy from light. Most leaves are also thin to reduce the distance that gases need to diffuse in and out from stomata.

Know >

1 What colour are parts of plants in which photosynthesis occurs?

2 Why are many plant leaves large?

3 Why are plant leaves also often very thin?

4 By what process do gases move into and out from a leaf?

Apply >>

5 Why are palisade cells full of chloroplasts?

6 Why does carbon dioxide diffuse into a leaf during the day?

7 Why does oxygen diffuse into your blood?

Extend >>>

8 What would happen if spongy cells in leaves did not have gaps between them?

9 What would happen to leaves without a cuticle?

» Core: Root structure

Almost all plants that grow on land have roots. These hold plants in the soil or on the rock or tree onto which they are growing. Some plants like large trees have deep roots that search out underground stores of water. Other plants like the grass in people's lawns have many shallower roots that absorb as much water as they can after a rain shower.

Key word
Fertilisers are chemicals containing minerals that plants need to build new tissues.

Plants also need to absorb small quantities of minerals from the soil. They need to absorb small amounts of magnesium to make chlorophyll and nitrates to make proteins for growth and repair. Nitrates are natural plant **fertilisers**. These are also absorbed through plant roots.

Root hair cells

These specialised cells have a small section that pokes out into the soil. This increases the surface area of the cell to absorb water from the soil. Whilst each individual root hair is often very small, their combined length in any individual plant is huge. The total length of all root hairs in a single wild oat plant is over 50 miles! This large surface area means plants are able to absorb significant volumes of water through their roots.

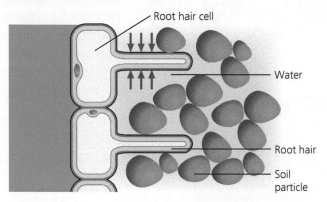

▲ Root hair cells stick out into the soil to absorb more water and minerals

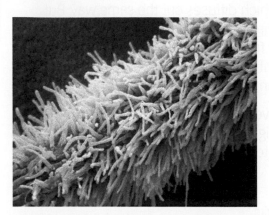

▲ Look how many root hairs are on this one tiny root

Xylem vessels and transpiration

Plants do not suck water from the soil. It moves into root hair cells because of a process called osmosis. This is a special type of diffusion when water moves from a higher concentration of water to a lower concentration across a membrane. Water moves from the root hair cells into **xylem** vessels. These are made from dead cells that have joined up to form long thin tubes. These vessels stretch from the roots to the leaves. Water flows up them.

So we can think of xylem vessels a little like our veins or arteries. But plants do not have a heart to pump water from their roots to their leaves. The world's largest tree is a giant redwood called General Sherman in the USA. It is nearly 84 m tall. So how does it, and all other plants, get water from their roots to their leaves against gravity?

The answer is a process called **transpiration**. Plants release water vapour from their stomata found on the bottom of their leaves. As this water evaporates from the leaves more water is 'pulled' upwards from the roots. There is a continuous flow of water up xylem vessels from the roots to the leaves. This is called the 'transpiration stream'. This process occurs continuously throughout the life of a plant.

▲ General Sherman: the world's largest tree

Amazingly plants are able to lose up to 90% of the water they absorb through their roots during the process of transpiration. They retain the 10% for use in photosynthesis. This shows how important it is to plants to keep water continuously streaming up to their leaves in transpiration.

Scientists have calculated that the maximum height any tree can reach is about 120 m tall. Beyond this height transpiration no longer works. Therefore the height of all plants on our planet is limited by this process.

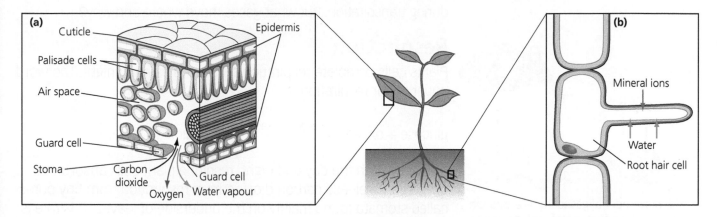

Phloem vessels and translocation

The water that reaches the leaves of plants is combined with carbon dioxide and converted into glucose and oxygen during photosynthesis. But what happens to the glucose that is produced? You learned earlier that it is also stored as starch in structures like potatoes and rice, and used to make cellulose for cell walls. It is used by all of the plant's cells for respiration to provide energy for the life processes. So the glucose made during photosynthesis must be moved to all cells of a plant. This process is called **translocation** and occurs in **phloem** vessels.

Phloem cells are alive, unlike those in xylem vessels. They form long thin tubes, like xylem, but have permeable ends which allow the glucose dissolved in water to move through. They take this sugary-water to all other parts of a plant.

Know >

1 Why do plants have roots?

2 What else is absorbed by roots besides water?

3 What are fertilisers?

4 Why do plants have root hair cells?

5 How are xylem cells adapted for their function?

Apply >>

6 How are xylem and phloem different?

7 What effect would it have on plants if they were unable to grow more xylem vessels?

Extend >>>

8 What are the similarities and differences between transpiration and translocation?

9 What would happen to plants if they didn't have root hair cells?

» Extend: Movement of gases in leaves

Gases move from the air into and out from the stomata mainly found on the underside of leaves. These tiny pores open and close to regulate the flow of carbon dioxide and oxygen for photosynthesis and respiration. They also regulate the evaporation of water vapour during transpiration. But what gases move where and why?

Respiration

Plants cells complete respiration during the day and night. The word equation for respiration is:

$$\text{glucose + oxygen} \xrightarrow{\text{energy released}} \text{carbon dioxide + water}$$

So during both the day and night, respiring plant cells absorb oxygen and release carbon dioxide from respiration from tiny pores called stomata found mainly on the underside of leaves. This rate is relatively constant throughout.

Photosynthesis

Plant cells complete photosynthesis during the day. This is the word equation for photosynthesis:

$$\text{carbon dioxide + water} \xrightarrow[\text{chlorophyll}]{\text{light energy}} \text{glucose + oxygen}$$

During the day, photosynthesising plant cells absorb carbon dioxide and release oxygen which both move through stomata. This means that during the day we can see overall movement of gases as a result of photosynthesis as shown in this diagram:

Overall

So we see the following movements of gases into and out from leaves during the day and night:

We can see that during the day plant cells require oxygen for respiration and make carbon dioxide as a waste product. During the day, they also require carbon dioxide for photosynthesis and produce oxygen as a waste product, so the movement of gases is relatively balanced.

At night plants still require oxygen for respiration and make carbon dioxide as a waste product. But they don't photosynthesise during the night, so they don't require carbon dioxide for photosynthesis nor

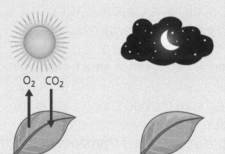
▲ Movement of gases from respiration into and out from a leaf during the day and night

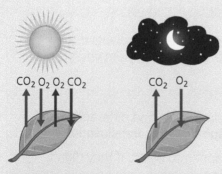
▲ Movement of gases from photo–synthesis into and out from a leaf during the day and night

▲ Overall movement of gases into and out from leaves during the day and night

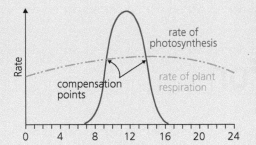

▲ The rate of photosynthesis and respiration during a day

produce oxygen as a waste product. So the movement of gases is not balanced.

Photosynthesis does not occur at the same rate throughout the day. You learned earlier that the more photosynthesis occurs at higher light intensities. So often, more photosynthesis occurs during the middle of the day. Here the rate of photosynthesis overtakes the rate of respiration. So there are therefore points during the morning and afternoon when the rates of both reactions are equal. These are called **compensation points**.

Are photosynthesis and respiration opposites?

It is easy to look at the equations for respiration and photosynthesis and think they are the reverse of each other. The products and reactants are overall. But these two crucial reactions are not the opposite in another key regard. Can you work it out?

The answer is **the flow of energy**. Light energy, often from the Sun, powers photosynthesis. This energy is stored in the form of glucose in plants. You can see an arrow showing the light energy flowing into the equation for photosynthesis. But energy is released from glucose during respiration. This energy allows the plant to complete the seven life processes needed to stay alive. You can see an arrow showing the energy flowing out from the equation for respiration.

But this release of energy does not just allow plants to live. Most organisms on our planet depend upon the plants at the bottom of food chains. The energy captured during photosynthesis flows from the Sun through plants to support almost all other life.

The only food chains on our planet that do not depend upon photosynthesising plants or algae are those found around deep sea volcanic vents. Here it is very hot and extremely dark so no plants could survive. Amazing bacteria have evolved to consume chemicals released from the vents. These bacteria are at the bottom level of short food chains. These are the only life forms on our planet to not rely upon photosynthesis.

▲ Deep sea thermal vents with their unique species of life

Tasks

1. What chemical reactions happen in plants during the day?
2. What chemical reaction only happens at night?
3. What gases move into a leaf during the day?
4. What gases move out from a leaf during the night?
5. What is the name given to the tiny pores on the undersides of leaves?
6. What is a compensation point?
7. What else is different about photosynthesis and respiration as well as their reactants and products?

Enquiry:
Is chlorophyll essential for photosynthesis?

You learned earlier in the chapter that the following are essential for photosynthesis:

- light energy
- water
- carbon dioxide
- chlorophyll.

We know that we kill plants if we put them in the dark. This may take a week or two whilst the plant uses it stores of energy to survive, but it will eventually die. The same is true of water. If we stop watering plants they will die. Some, like cacti, can survive for long periods without water but all eventually die. It is difficult to remove carbon dioxide from the atmosphere surrounding a plant. But if we did, we would see the plant die. It cannot photosynthesise without carbon dioxide and so would die. But what about chlorophyll? How can we prove that plants need this for photosynthesis? We will complete an experiment to prove this.

Some plants like geraniums have variegated leaves. These have often been grown by gardeners for the pretty patterns in their leaves. Parts of variegated leaves are green. The cells in these places have chloroplasts containing chlorophyll and so are coloured green. Other parts of the leaves are white. Here these cells do not possess chlorophyll and so are not green. We will use a variegated leaf to test whether photosynthesis occurs in the white parts as well as those green sections containing chlorophyll. By doing so, we will see if chlorophyll is essential for photosynthesis.

▲ The variegated leaf of a geranium plant

Aim

To prove that chlorophyll is essential for photosynthesis.

Equipment

Variegated leaf, boiling tube, water, ethanol, water bath set to 80°C, timer, white tile, iodine solution

Method

1 Draw an image of the variegated leaf showing which sections are green and which are white.

2 Boil a plant leaf in a boiling tube of water for one minute. This stops all of the chemical reactions happening.

3 Transfer the leaf to a boiling tube of ethanol (alcohol) in a water bath at 80°C for 5 minutes. This removes the chlorophyll from the leaf.

4 Wash the leaf in water and spread onto a white tile.

5 Apply iodine solution to the whole leaf.

6 Redraw the leaf showing which parts have been stained brown and which parts are now black.

Health and safety

Ethanol is an alcohol and so it is flammable. It will boil at a lower point than water (78°C). It should not be put near a flame. Safety goggles must be worn at all times.

▲ A variegated leaf before and after the experiment

Results

The diagrams you have drawn should look like the images in the photograph on the left.

The regions of the leaf that were green should now have turned black. The regions of the leaf that were white should remain white or be stained brown by the iodine.

Conclusions

When plants photosynthesise, they produce glucose. This is stored as starch in the photosynthesising cells. When iodine is placed upon them, it reacts with the starch turning that part of the leaf black.

The white cells of the variegated leaf did not turn black. They remained white or were stained brown by the iodine, but they did not change colour. This proves that they were not photosynthesising. This in turn proves that chlorophyll is required for photosynthesis.

We test for starch in this experiment not glucose, which is the product of photosynthesis. This can be confusing. Why do we test for starch instead of glucose? The answer is that using iodine to test for starch is much easier than using Benedict's reagent to test for glucose.

Key fact

Iodine is used to test for the presence of starch.

❶ What is used to test for starch?
❷ What colour will it turn in the presence of starch?
❸ What is used to test for glucose?
❹ Why do gardeners grow plants with variegated leaves?
❺ Why do we test for starch not glucose in this experiment?

Genes

Learning objectives

19 Evolution

In this chapter you will learn...

Knowledge

- that evolution is explained by natural selection
- that biodiversity is vital to stop extinctions
- that variation helps species adapt to environment changes
- that ecosystems with many species have more resources available for other populations
- the definitions of the terms pollution, natural selection, extinct, biodiversity, competition and evolution

Application

- how to use evidence to explain why a species has become extinct or survived changing conditions
- how to suggest whether evidence for a species changing over time supports natural selection
- how to explain how a lack of biodiversity can affect an ecosystem
- how to describe how preserving biodiversity can provide useful products and services for humans

Extension

- how to suggest how natural selection can lead to changes in a population
- how to explain how evolutionary changes may have occurred
- how to evaluate ways of preserving plant or animal material for future generations

20 Inheritance

In this chapter you will learn...

Knowledge

- that inherited characteristics are the result of DNA transfer from parents to offspring during reproduction
- that chromosomes are long pieces of DNA containing many genes
- that the DNA of every individual is different, except for that of identical twins
- that there is more than one version of each gene
- the definitions of the terms inherited characteristics, DNA, chromosomes and gene

Application

- how to identify DNA, chromosomes and genes in a diagram
- how to use a diagram to show how genes are inherited
- how to explain how mutations can affect organisms
- how to explain why offspring from the same parents look similar but are not usually identical

Extension

- how to suggest arguments for and against genetic modification
- how to suggest benefits from scientists knowing all the genes in the human genome
- how to describe how the number of chromosomes changes during cell division, production of sex cells and fertilisation
- how to explain why scientists Watson, Crick and Franklin were so important

» Transition: Adaptations

Adaptations are ways in which organisms are better suited to their environment. All living organisms are adapted to the environment in which they live. When we describe adaptations, it is important that they have an explanation of how they help the organism. The fact that polar bears are white is not an adaptation. It is just a fact. The fact that polar bears have white fur so they are camouflaged and can hunt more effectively is an adaptation.

All organisms have evolved certain adaptations. You will learn more about evolution later in this chapter.

A cold environment

Polar bears have small ears. This means that they do not lose as much heat in their cold environment. Animals that live in the hot environments like desert foxes often have large ears to lose more heat. Polar bears have strong, muscular legs for swimming and running to help them catch seals as prey. They have a thick layer of fat under their fur to act as insulation to keep them warm. The fat is also a store of energy allowing them to survive the cold winter.

Pine trees have thick bark to protect them from the cold temperatures. They also have pine cones which protect their seeds during the cold winter months. Instead of large leaves, they possess small, thin, waxy needles which reduce water loss for when most water is frozen. They keep their needles during the winter to maximise photosynthesis.

A hot environment

Camels can withstand much higher temperatures than other animals so they do not need to sweat to keep cool. This means they do not lose as much water as other animals. Camels have humps which contain fat which are a store of energy. However, they don't have as much fat in other parts of their bodies which keeps them as cool as possible. Camels also have bushy eyelashes and can close their nostrils to keep the sand from their eyes and nose.

Your knowledge objectives

In this chapter you will learn:
- that evolution is explained by natural selection
- that biodiversity is vital to stop extinctions
- that variation helps species adapt to environment changes
- that ecosystems with many species have more resources available for other populations
- the definitions of the terms pollution, natural selection, extinct, biodiversity, competition and evolution

See page 225 for the full learning objectives.

▲ How many adaptations of polar bears can you see in this picture?

▲ How many adaptations of pine trees can you see in this picture?

◄ How many adaptations of camels can you see in this picture?

▲ How many adaptations of cacti can you see in this picture?

Instead of large leaves, cacti possess needles which reduce water loss. This is very important in a desert when rain is rare. The spines also protect the cacti from being eaten by herbivores after their water. Some cacti have shallow roots that spread to catch as much rain water as possible. Others have fewer, deeper roots which tap into underground water sources.

An aquatic environment

Fish possess gills which absorb oxygen from water. Many fish have a streamlined shape to make it easier for them to swim quickly through the water. Predatory fish need to swim fast to catch prey and those that are preyed upon need to swim fast to avoid being eaten.

Water lilies have leaves that sit on the surface of water to maximise photosynthesis. The roots sink into the riverbed to keep them in one place. So they have thick, flexible stems to mean they can bend and move with the water current without breaking.

▲ How many adaptations of fish can you see in this picture?

▲ How many adaptations of water lilies can you see in this picture?

Know >

1 How are polar bears' ears adapted?

2 Why do pine trees have thick bark?

3 How are camel's eyes adapted?

Apply >>

4 What size ears might animals in the desert have and why?

5 Why are coconut trees flexible?

6 Why do water lilies have flowers that float on the surface?

Enquiry >>>

7 What other adaptations of animals that live in hot and cold conditions can you find out?

» Core: The theory of evolution by natural selection

The theory of **evolution** explains how all living species alive today originally came from a common ancestor that we may all be related to. We don't know very much about this species but we call it the **last universal common ancestor**. We think that it lived over three and a half billion years ago.

The process of evolution

Evolution can be separated into the following steps:

1 In every population of organisms there is variation.

All organisms within any species of life on Earth have small differences between them, which we call variation. How do lions vary? Some lions will have sharper claws, stronger muscles and better eye sight. Some will have bigger manes or louder roars. It is sometimes easy to forget that all life has evolved, not just animals. So how might individuals within a population of oak tree vary? Well some will be faster growing to reach the light for photosynthesis. Others might produce leaves earlier in spring for the same reason.

Because the theory of evolution explains how all life on Earth has evolved, we can see the variation in all populations of organisms. Choose any one and try to work out what the differences are.

We call differences within populations **biodiversity**. You will learn more about this later in this chapter.

2 This must mean that some organisms are better adapted than others.

Think about the populations above. What makes a better adapted lion? The biggest lion, with the stronger muscles, fastest speed and strongest claws and teeth is probably the best adapted.

3 These organisms are more likely to have offspring.

Within any population some organisms are going to be more successful. We humans have a slightly different sense of what successful is when compared with other organisms. For the vast majority of life on Earth, just living day to day is a struggle. They face **competition** from other individuals within their population for food and other resources.

The organisms that are better adapted are more likely to survive each day. This means they are more likely to grow to reach

▲ A well-adapted lion

Key words

Evolution is the theory that the animal and plant species living today descended from species that existed in the past.

A **population** is a group of the same species living in an area

Biodiversity is the variety of living things. It is measured as the differences between individuals of the same species, or the number of different species in an ecosystem.

Competition is when two or more living things struggle against each other to get the same resource.

reproductive maturity. This is the point that they are able to breed and have offspring.

Badly adapted organisms are much less likely to have offspring. This means that they are less likely to pass on their poorly adapted genes.

4 Their offspring are more likely to have their parent's advantageous characteristics.

We are a little like our parents. We might be as tall as our fathers, or have our mother's hair. Our noses might be most similar to our fathers, or our eye colour the same as our mothers. This is called **inheritance**. Because of this process, successful parents are likely to have offspring with the same adaptations.

This means that the offspring of the biggest lion, with the stronger muscles, fastest speed and strongest claws and teeth, are likely to be like their parents. So they too are likely to have the stronger muscles, fastest speed and strongest claws and teeth.

▲ Inheritance of characteristics is seen in this family photograph

5 Evolution occurs when these steps are repeated over many generations.

Scientists think that life has existed on Earth for over three and a half billion years. During this time, the theory of evolution says that all species have been slowly improving. The best adapted individuals in every population have been most likely to have offspring with their characteristics. This has meant that slowly, generation by generation, all populations have become better adapted to their environment. They have evolved.

As well as this though, the reverse of is also true. The least most poorly adapted individuals in any population are least likely to have offspring. This means that their adaptations will not be passed on. This can happen to a whole population, not just individuals within it. When this does it becomes **extinct**.

Key words

Extinct is when no more individuals of a species remain.

A **species** is a group of organisms that can interbreed to have fertile offspring.

From one to millions of species?

We don't know for sure that there was one **species** of life called our **last universal common ancestor** over three and a half billion years ago. We don't know much about what this organism did or how it lived. We do know that all species of life on Earth seem to have evolved from it.

Scientists estimate that there are around 8.7 million species of life on Earth at the moment. It is highly likely that all these species of life have a genetic template made from DNA. It is likely that they have all evolved from our **last universal common ancestor** which we assume had genetic information made from DNA too.

Know >

1 What is evolution?

2 What term describes differences within populations?

3 Why is it easier to see evolution in bacteria than in humans?

4 Why did the dinosaurs become extinct?

Apply >>

5 Why is variation important for evolution?

6 What is likely to happen to badly adapted species?

Extend >>>

7 What can you find out about our last universal common ancestor?

8 How is the evolution of horses' feet linked to the hardness of the ground?

>> Core: Charles Darwin

You learned on the previous page that evolution is the slow process of improving individuals within a population over many generations by the natural selection of the most adapted as the most likely to have offspring. This process has been repeated many times as populations become better adapted. This is evolution.

This seems like a very sensible idea in today's society and the vast majority of scientists alive today believe in it. This was not always the case however. Around 200 years ago, the Church was a more powerful organisation than today. Until this point, the vast majority of people believed that God created the universe and all life within it. This is called **creationism**. Some people still believe this today.

Charles Darwin and the Galapagos Islands

Charles Darwin lived between 1809 and 1882 and was an English naturalist and geologist. He is best known for his theory of evolution, which he co-published with Alfred Wallace in 1859.

After university and against his father's wishes, Charles Darwin took a five-year voyage around the World on a ship called the Beagle. He first sailed to South America visiting Brazil and Chile, before stopping on the Galapagos Islands many miles from the coast of Ecuador.

The Galapagos Islands are a small group of islands formed by the tops of underwater volcanoes poking out of the Pacific Ocean. Some of these islands are still growing as more underwater magma comes from the volcano underneath.

▲ Charles Darwin, the father of evolution

Importantly, the Galapagos Islands have unique differences. Some are small, flat and dry with many cacti. Others are bigger and taller, with more vegetation. These different ecosystems allow a hugely diverse group of animals and plants to live there including many that can be found only there.

Darwin travelled over the Galapagos Islands, taking notes, drawing pictures and collecting live animals and plants. After his ship left the Galapagos Islands and continued on its voyage, Darwin had plenty of time to think about what he had seen.

Darwin's finches

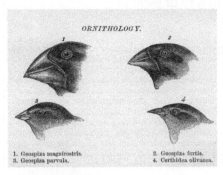

ORNITHOLOGY.

1. Geospiza magnirostris.
3. Geospiza parvula.
2. Geospiza fortis.
4. Certhidea olivacea.

▲ Darwin's finches

Charles Darwin was particularly interested in the different finches he saw on the Galapagos Islands. He saw similarities between them which suggested they may all have once been part of the same population. He also saw differences in the way they had adapted to the conditions on the island they were found on. From this Darwin developed his theory of evolution.

A key step in this theory is that the best adapted organism within a population is most likely to survive and reach reproductive maturity. Darwin called this process **natural selection**. He meant that nature was 'selecting' the best adapted. He also called this '**survival of the fittest**'.

Key word

Natural selection is the process by which species change over time in response to environmental changes and competition for resources.

Fear of the Church

Darwin returned to England in 1836 and became a famous scientist because of the animals and plants he had discovered on his trip. He did not publish his developing theory of evolution though. He was worried about the reaction from the Church which was very powerful at the time.

Twenty years later Darwin was sent a report by Alfred Wallace who had independently developed the theory of evolution on his own. Darwin and Wallace agreed to jointly publish their work.

▲ In 1871 this cartoon was published showing Darwin's head on an ape's body

Evolutionary tree

You learned earlier in this chapter that scientists now think there are 8.7 million species of life on Earth and they are likely to have evolved from our **last universal common ancestor** over three and a half billion years ago. Charles Darwin was the first person to write about this process.

▲ In 1837 Charles Darwin drew the first evolutionary tree and wrote "I think" above it

Evolutionary trees are diagrams that show how different species evolved. They are a little like family trees for different species. In 1837 Darwin drew the first evolutionary tree and wrote "I think" above it. He wrote:

"Therefore I should infer from analogy that probably all the organic being which have ever lived on this earth have descended from one primordial form, into which life was first breathed."

Know >

1 Which scientist developed the theory of evolution?

2 Which islands did Darwin visit?

3 Why did Darwin wait before publishing his work?

4 What is survival of the fittest?

5 What is natural selection?

Apply >>

6 Why did the finches on the Galapagos Islands help Darwin?

7 Charles Darwin also studied geology (rocks) and so was familiar with fossils. How might this have helped him with developing this theory of evolution?

Extend >>>

8 Use the internet to find an evolutionary tree for vertebrates.

>> Core: Biodiversity

Biodiversity is a measure of the differences between organisms within a species or the total number of species in an ecosystem. So it is the differences between organisms in any one population.

Look around your classroom or family dinner table. What biodiversity can you see? You might see differences in height, weight, skin and hair colour, piercings, scars and tattoos. All these examples of **variation** are also examples of biodiversity.

But biodiversity is also a measure of how much variation there is in the species within an ecosystem. Let's compare the Artic or a desert with the rainforests. Which one is the most biodiverse? It is the rainforest. Here thousands more species of animals and plants are found.

Here we will consider the difficulties that ecosystems with lowered biodiversity have. The difficulties that populations with lower variation have are covered in Pupil's Book 1, Chapter 19

> **Key word**
>
>
>
> **Variation** is the differences within and between species.

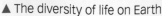

▲ The diversity of life on Earth

▲ We have developed the aspirin drug from the bark of the willow tree

The importance of biodiversity in ecosystems

Biodiversity is extremely important when we consider ecosystems all over our planet. Scientists think there are around 8.7 million species of life, but we have only formally identified around 1.2 million of them. Who knows how useful some of these other millions could be. Is the cure for cancer or HIV/AIDS waiting for us in one of them? What new resources or technologies can we develop from looking at these populations?

Variation within a population of organisms also helps it evolve. If all organisms in a population were identical then a change in environmental conditions, a new type of pollution or a new predator could wipe it out. If variation exists within this population, then some individual organisms are likely to be better adapted. These can then evolve to counter the change.

Human impact on biodiversity

A reduction in biodiversity makes individual populations more likely to become extinct. It also makes is less likely we will find new drugs, resources or technologies.

Preserving biodiversity for the future

In case the rate of extinction continues, scientists have begun to store genetic material to help restore future populations. This genetic information is in the form of frozen sperm and eggs from animals, and either by freezing tissues or seeds from plants.

▲ The Svalbard Global Seed Vault contains over 10,000 seed samples

These vital biological samples are stored in relatively safe places. Perhaps the most well-known of these is the Svalbard Global Seed Vault built in the ice of a Norwegian island.

Know >

1 What is biodiversity?

2 About how many species of life exist on Earth?

Apply >>

3 What impact have humans had on biodiversity in many ecosystems?

4 How are we preserving biodiversity for future generations?

Extend >>>

5 What will the consequences be if we continue polluting our air, water and land?

» Extend: Evidence for evolution

Scientific evidence is gathered to support or disprove scientific theories. This is key part of science, for it allows us the become more (or less) confident in the latest scientific theory. A scientific theory is an explanation that is based upon facts that have been repeatedly confirmed by experiments or observations.

The theory of evolution is now thought of by many as both a theory and a fact. The evidence to support the theory of evolution is good enough for nearly all scientists to support it as a fact.

The peppered moth

The Industrial Revolution occurred between about 1760 and 1830. During this time the development of steam power meant that number of machines making products increased significantly. No longer were many items hand made, but they were now produced in significantly larger numbers in new factories.

This did help the standard of living for many people, but also gave rise to a huge increase in air pollution. This actually changed the colour of the bark of many trees from pale to a darker colour.

Before the Industrial Revolution, the majority of peppered moths were a light colour and only a very small proportion were darker. This is a form of variation. At this point the darker moths were easier to spot by birds and so more often eaten.

When the trunks of the trees turned darker because of the pollution the lighter coloured moths no longer had the evolutionary advantage. They were now the ones that stood out and were more likely to be eaten.

The proportion of the population of moths that were black increased as they were now much better adapted to their environment.

This provides evidence for both the importance of biodiversity in populations and evolution.

▲ Changes in colour of the peppered moth provide evidence for evolution

The pentadactyl limb

Pentadactyl means having five digits on each limb. We see this pattern in our own hands and feet. However, many other vertebrates (animals with backbones like us) have a very similar structure as well. This is sometimes difficult to see.

Examples of organisms with the pentadactyl limb include amphibians like frogs, reptiles like lizards, birds, and other mammals like the horse, whale and bat.

Similarities in the patterns of these bone structures strongly suggests a common ancestor of all these organisms that had five digits on its limbs. It therefore provides evidence for evolution.

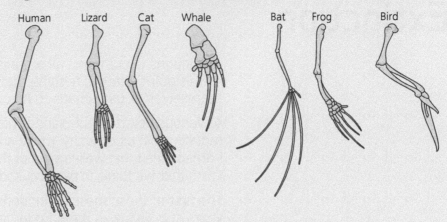

Human Lizard Cat Whale Bat Frog Bird

▲ The similarities in the pentadactyl limb provides evidence for evolution

Antibiotic-resistant bacteria

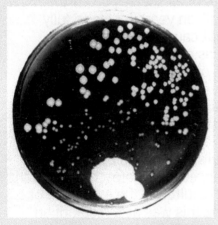

▲ Alexander Fleming's original Petri dish in which he discovered the first antibiotic

Sir Alexander Fleming (1882–1955) was a Scottish biologist. He accidentally discovered the first antibiotic, called penicillin, in 1928 for which he was awarded the Nobel Prize. This discovery has saved hundreds of millions of lives.

At the time that Fleming discovered penicillin it killed all known bacteria. It has only taken several decades for the bacteria to fight back. Relatively quickly they have evolved a resistance to our antibiotics. In the last few years we have discovered bacteria that are now resistant to all our antibiotics. The NHS have recently said that antibiotic resistance is one of the most significant threats to patients' safety in Europe.

Because the life cycle of bacteria occurs in hours not years, we are able to see the evolution of antibiotic-resistance in bacteria in front of our very eyes. This provides evidence for evolution.

Tasks

1. What does the peppered moth provide evidence for?
2. What environmental change caused the change in the peppered moth population?
3. In what animals do we see the pentadactyl limb?
4. What does the pentadactyl limb suggest?
5. How does antibiotic resistant bacteria provide evidence for evolution?

Enquiry:
Extinction

Extinctions occur when all organisms within a species die out. At this point, there is nothing that can be done by the organisms themselves or us humans to help.

At various points in our planet's history we have seen vast reductions in biodiversity. These are called **mass extinctions**. Perhaps the most well-known of these is when an asteroid hit Earth that we think, in part, caused the extinction of the dinosaurs.

The first of these mass extinctions was around 450 million years ago. This was the third largest extinction event in history. We think this was caused by changes to the Earth's climate. At this point the majority of life still lived in the oceans and seas. Around 375 million years ago, the second mass extinction occurred. Amazingly, three quarters of all species on Earth became extinct. It took over 100 million years for life to fully recover. We are not really sure what caused this extinction event. The third mass extinction was even worse. It occurred 250 million years ago. Around 96% of all species on Earth became extinct and all species alive today have evolved from the 4% that were not killed. Scientists think this was caused by the impact of a collision with an asteroid. The next extinction occurred around 200 million years ago. Fewer species became extinct during this event. We think it was caused by an asteroid collision, volcanic eruptions or changes in climate. The fifth event occurred 75 million years ago and resulted in the extinction of the dinosaurs. Many scientists think this was as a result of an asteroid collision that formed a crater off the Mexican coast.

▲ Did a collision like this cause mass extinctions?

It is important to remember that after all of these extinction events an opportunity arises for new organisms to evolve. After the last extinction event we see many more mammals evolving including the development of horses, whales, bats and primates (including us). Birds, fish and possibly reptiles also grew in number.

The latest mass extinction event

For a long time Lonesome George was the last individual Pinta Island tortoise left alive. He lived on the Galapagos Islands for over one hundred years and for most of his life he was alone so unable to mate. Sadly, he died in 2012 and this is now yet another species that has become extinct.

A similar situation exists with the northern white rhino. There are only three northern white rhinos left, one male and two females, and they are all too old to breed anymore. Just like Lonesome George, they cannot reproduce. When they die, yet another species will be extinct.

▲ This was Lonesome George, the last ever Pinta Island tortoise

But why are we in this position? The rest of Lonesome George's species died when humans introduced goats onto Pinta Island. The goats became an invasive species. You learned about this in Pupil's Book 1, Chapter 17. The goats quickly ate many of the plants that the tortoises needed to survive. Only Lonesome George remained. We humans are responsible for this extinction. Similarly, we have killed nearly all of the other northern white rhinos. Some were shot or trapped for sport. Others were killed for the horns which are sold as medicines that we know don't work.

Many scientists believe we are currently going through the sixth of these mass extinction events. Sadly, this is the first and only one that has been caused by one species. Can you guess which species it is? It is us humans. The World Wildlife Fund believe that the current rate of extinction or species per year is between one thousand and ten thousand times more than it would normally be. This means between 0.01% and 0.1% of all species become extinct each year. This is a frightening thought.

▲ The last male northern white rhino lives under armed guard (in 2017)

1. What are mass extinctions?
2. What proportion of species were killed during the worst mass extinction event?
3. What happened in the years after a mass extinction event?
4. What is an invasive species?
5. Why would an extinction event occur if an asteroid hit an ocean?
6. Why would an extinction event occur if an asteroid hit the land?

Your knowledge objectives

In this chapter you will learn:
- that inherited characteristics are the result of DNA transfer from parents to offspring during reproduction
- that chromosomes are long pieces of DNA containing many genes
- that the DNA of every individual is different, except for that of identical twins
- that there is more than one version of each gene
- the definitions of the terms inherited characteristics, DNA, chromosomes and gene

See page 225 for the full learning objectives.

▲ The trapdoor spider

▲ The skull of a fox shows eyes facing forwards

» Transition: Predators and prey

For millions of years, predators have hunted and killed prey. For equally as long prey have tried to avoid being hunted and killed. You learned in the previous chapter that adaptations are ways in which organisms are better suited to their environment. All organisms inherited these adaptations from their parents. You will learn more about inheritance later in this chapter.

What adaptations do predators inherit?

All predators need to be able to catch prey. Some like the cheetah chase directly after prey and hope to be faster than them over short distances. These predators are built for speed. They often have large muscles which can explosively generate a lot of speed very quickly. They also often have very quick reactions.

Other predators, like wolves, chase prey more slowly but for much longer times, often catching them when they are close to death from exhaustion. These predators are built for stamina not pure speed. Of course they need to be fast enough to keep up with their prey, but they chase them for hours or even days so can be a little slower. These and some other species have evolved to hunt in packs. Other predators like polar bears hunt alone.

Other predators wait patiently, without moving, and catch their prey by surprise. The trapdoor spider digs a small burrow and covers it with a trapdoor. It makes silk tripwires around the trapdoor. It then waits until its prey stands on the trapdoor or catches one of its silk tripwires. It quickly bursts out of its trapdoor and catches its prey. Other predators like lizards and snakes don't build traps, but reply upon camouflage to fool their prey. They remain very still as their prey moves towards them.

Predators have sharp teeth and claws to hold onto or kill their prey when they have caught it. Other predators like scorpions have developed stings to stun their prey.

They also have eyes to the front of their heads, not on the side. This means they can judge distance between themselves and their prey much easier. We humans have eyes like this. For many years we have been predators too.

What adaptations do prey inherit?

▲ The skull of a rabbit shows the eyes to the side

All prey need to be able to run away from their predators. Prey like rabbits and deer are very fast and can often change direction very quickly. This helps them avoid predators that are very close to them. So prey often have large muscular legs like the predators that are chasing them.

Unlike predators, prey are often found in large numbers. Flocks of birds can have millions of individuals. Wildebeest and other African animals migrate through the grassland plains in populations of hundreds of thousands. Some smaller fish like sardines that are eaten by larger fish live in shoals of similar sizes. When a predator attacks such a large population it is often difficult for it to identify one individual to catch. Often the confusion means no prey are caught.

Some prey like frogs and toads have developed poisons as chemical defences. You learned about the poisonous glands on the back of the cane toad in Pupil's Book 1, Chapter 17.

Like predators, some prey are camouflaged. Seahorses blend in with the coral they live amongst. Stick insects look very much like sticks to hide from predators. Other prey have bright colours to warn off predators. In nature, these colours often signify poison. The harmless Mexican milk snake has evolved to copy the red, black and yellow markings of the Texas coral snake.

► Which is the poisonous coral snake?

Unlike predators, prey usually have their eyes on the sides of their head. This gives them a much wider field of view. This means they can see predators more easily.

Know >

1 How are cheetahs adapted?

2 What defence has the cane toad evolved?

Apply >>

3 Why do wolves hunt in packs?

4 Why are predator's eyes usually on the front of their heads?

5 Why are prey's eyes usually on the sides of their heads?

Enquiry >>>

6 Cover one of your eyes, extend your arms and point one finger on each hand toward each other. Try to touch your two fingers together with one eye closed. It is much harder than with both open. What does this prove?

DNA is a molecule found in the nucleus of cells that contains genetic information.

The **genome** is one copy of all of an organism's DNA, found in every diploid body cell.

Sperm and eggs in animals (and pollen and eggs in plants) are **haploid** cells which contain half an organism's DNA.

All cells except sperm and eggs are **diploid** cells which contain all an organism's DNA (its genome).

Chromosomes are thread-like structures containing tightly coiled DNA.

» Core: From the genome to DNA

You are you because of your genetic code which is made from **DNA**. This was fixed the moment your father's sperm fertilised your mother's egg. At this point the DNA from both sex cells came together to form you. If you are an identical twin, you share your genetic code with your twin. If you are not, your genetic code is unique. Your **genome** is one copy of all of your genetic code. This is found in almost all of your cells.

Haploid sex cells are sperm, eggs and pollen. These have half the genetic information of diploid body cells. These need only to have half of an individual's genetic information, so they can fuse with another haploid cell during fertilisation to form a **diploid** fertilised egg cell. This will grow into a baby (or new plant) with a new, unique genome.

Inside all of your cells except your red blood cells is a nucleus. Red blood cells don't have a nucleus to maximise the volume that can absorb oxygen. You learned about adaptions of cells in Pupil's Book 1, Chapter 16.

Inside the nucleus of diploid body cells is approximately 2 m of DNA. This is a long length to fit into cells that we cannot usually see without using a microscope. DNA is therefore tightly coiled into **chromosomes**. There are twenty-three pairs of chromosomes in each diploid body cell. We say they come in pairs, because you inherited twenty-three in your mother's egg and twenty-three in your father's sperm.

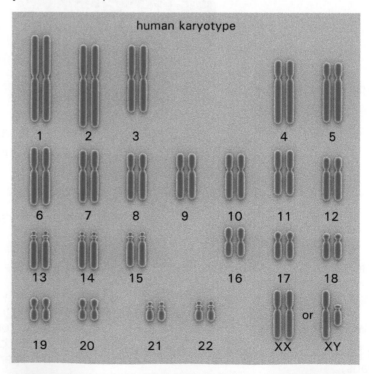

▲ Twenty-three pairs of chromosomes make up a person's genome

For most of a cell's life it's chromosomes are not tightly coiled up. When the cell is about to divide after it has copied itself, the twenty-three pairs of chromosomes coil themselves into characteristic x-shapes.

Other organisms have different numbers of chromosomes. The hedgehog has forty-five pairs of chromosomes and the pineapple has twenty-five.

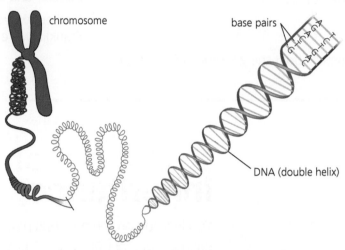

chromosome

base pairs

DNA (double helix)

▲ Chromosomes are made from coiled up sections of DNA

Key words

A **gene** is a section of DNA that determines an inherited characteristic.

Alleles are alternative forms of the same gene.

A **gene** is a section of chromosome made from DNA that determines an inherited characteristic. It does this by providing the code to make a protein. Humans have about twenty thousand genes. We inherit one copy of each gene from the sperm and another copy from the egg. These alternative copies of the gene are called **alleles**. So you have inherited alleles for eye colour, blood group and all other examples of genetic variation you learned about in Pupil's Book 1, Chapter 19.

Genes and chromosomes are both made from DNA. This important biological molecule has a distinctive shape called a double helix. This was discovered by British scientist Francis Crick (1916–2004) and American James Watson (1928–) in 1953. They were awarded the Nobel Prize in 1962. Their work used important X-ray images taken by British scientist Rosalind Franklin (1920–1958), who died before she could be awarded the Nobel Prize.

DNA is made from four bases which fit together into base pairs. They are:

Key facts

The DNA of every individual is different, except for identical twins.
There is more than one version of each gene, e.g. different blood groups.

- adenine and thymine (A and T)

- thymine and adenine (T and A)

- cytosine and guanine (C and G)

- guanine and cytosine (G and C).

There are over 3 billion base pairs in the human genome.

Know >

1 What is your genetic code made from?

2 What is your genome?

3 How many pairs of chromosomes do diploid human cells have?

4 What are alleles?

Apply >>

5 What is unusual about the genomes of identical twins?

6 What is the difference between sperm and brain, liver or muscle cells?

Enquiry >>>>

7 How many chromosomes do other animals and plants have? Can you find a link between this number and any other factor?

8 Find out why scientists Watson, Crick and Franklin were so important.

>> Core: Monohybrid inheritance

You learned on the previous page that we inherit two copies of each gene; one from your mother and one from your father. Alleles are two copies of each gene. So you have alleles for your blood group, eye colour and all other **inherited characteristics**.

Inheritance of eye colour

Eye colour is a common example of an inherited characteristic. Here we only consider blue and brown eyes, not green. We use letters to simplify the patterns of inheritance. These are called the **genotype**. The physical characteristics of a genotype are called the **phenotype**.

Genotypes always come in pairs; one letter denotes a gene from your mother and a second the different copy of the gene from your father. Capital letters denote a **dominant** gene and lower case letters mean a gene that can be dominated. We call this a **recessive** gene.

Brown eyes are dominant over blue eyes which are recessive. So there are three possible genotypes for eye colour:

BB means a dominant brown eyed gene was inherited from both parents. This is called a **homozygous dominant** genotype. The phenotype of these alleles is brown eyes.

bb means a recessive blue eyed gene was inherited from both parents. This is called a **homozygous recessive** genotype. The phenotype of these alleles is blue eyes.

Bb means a dominant brown gene was inherited from one parent and a recessive blue eyed gene was inherited from the other parent. We call this a **heterozygous** genotype. Because brown is dominant over blue, the phenotype of these alleles is brown eyes.

Key words

Inherited characteristics are features that are passed from parents to their offspring.

The **genotype** is the genetic make-up of an organism, given as a pair of letters to denote alleles of a gene.

The **phenotype** is the physical characteristics of an organism's genotype.

A **dominant** characteristic is one that is observed even if only one allele is present.

A **recessive** characteristic is one that is only observed if no dominant alleles are present.

Homozygous dominant is the presence of two dominant alleles.

Homozygous recessive is the presence of two recessive alleles.

Heterozygous is the presence of one dominant and one recessive allele.

We show the possible probabilities of inheriting all three possible genotypes in a Punnett square. This is a special grid shaped like this:

The genotypes of the mother and father go into the top and left sections outlined in red. So if a brown eyed mother with genotype Bb and father with blue eyes and genotype bb would be filled in like this:

	B	b
b		
b		

Each of the four spaces here outlined in black is a possible genotype of an offspring. These four boxes are filled in from the letter above them and the letter to their left. So the first box will be Bb.

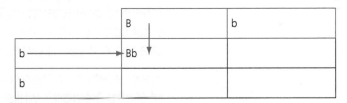

Follow the same pattern to fill in the other three boxes. This means the box below is also Bb and the two boxes to the right are both bb. So the competed Punnet square is as follows:

	B	b
b	Bb	bb
b	Bb	bb

It is important you analyse the probability of the outcomes of these genetic crosses. These are often given as percentages. The four boxes give the percentage probability of any individual child inheriting the given genotypes. It is important that you analyse the probability of both genotypes and then phenotypes.

▲ Blue eyes are given by the genotype bb

Genotype	Description	Percentage
BB	Homozygous dominant	0%
Bb	Heterozygous	50%
bb	Homozygous recessive	50%

Phenotype	Percentage
Brown eyes	50%
Blue eyes	50%

Other examples of monohybrid inheritance

Your ability to roll your tongue is another example of monohybrid inheritance for which you can use Punnett squares to determine probabilities of inheritance. Here we use the letter T.

TT means a dominant tongue rolling gene was inherited from both parents. This is called homozygous dominant. The phenotype of these alleles is tongue rolling.

tt means a recessive non-tongue rolling gene was inherited from both parents. This is called homozygous recessive. The phenotype of these alleles is non-tongue rolling.

Tt means a dominant tongue rolling gene was inherited from one parent and a recessive non-tongue rolling gene was inherited from the other parent. Because tongue rolling is dominant over non-tongue rolling, the phenotype of these alleles is tongue rolling.

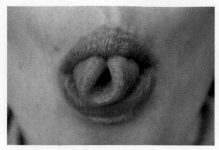

▲ The ability to roll your tongue is given by genotypes TT or Tt

How sex was determined

The first twenty-two pairs of chromosomes control many different inherited characteristics. The twenty-third pair specifically control what sex we are. Remember we can chose our genders, which can change, but our sex was determined at fertilisation.

We give letters to these chromosomes. We say all eggs are X. Half of sperm are also X whilst the remaining half are Y. If an X sperm fertilised an X egg then XX will become a baby girl. If a Y sperm fertilised an X egg than XY will grow into a baby boy.

Because the likelihood of having a baby boy or girl is 50% the proportion of male and female people alive is always very close to this.

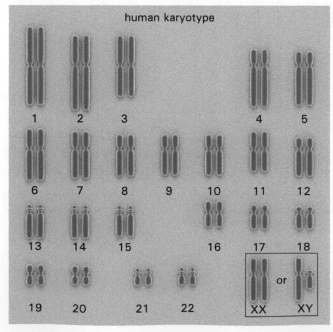

human karyotype

◄ The twenty third pair of chromosomes are the sex chromosomes

Know >

1 What is the genotype of an organism?

2 What is the phenotype of an organism?

3 How is sex determined in humans?

Apply >>

4 What is the homozygous recessive genotype for eye colour?

5 What is the homozygous dominant genotype for tongue rolling?

Extend >>>

6 Why is the ratio of men and women not exactly 1:1?

>> Core: Mutations and their effects

> **Key word**
>
> **Mutation** is changing of the structure of a gene.

Earlier in this chapter you learned that a gene is a section of DNA that determines an inherited characteristic. You also learned that a human is made from about twenty thousand genes. A **mutation** is a permanent change to a gene.

We tend to think of all mutations as bad and some mutation are. However, many mutations have no effect at all and others give organisms an evolutionary advantage. You learned about Darwin's theory of evolution by natural selection in Chapter 19.

Causes of mutation

Some mutations spontaneously occur. This means they occur without any reason at all. Other mutations are caused by exposure to specific environmental factors. These can be chemical substances called **carcinogens** or ultraviolet light, often from the Sun. Tar is one of 43 carcinogens found in cigarette smoke.

Some mutations lead to cancer. This occurs when a cell starts to copy itself many times over. This replication is out of control and causes a lump called a tumour. There are two types of tumour:

▲ The Sun's UV rays cause sunburn which can lead to skin cancer

Malignant: These are cancerous and are made of cells growing out of control. They can invade nearby tissues and spread.

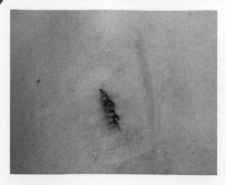
▲ Stitches after an operation to remove a skin cancer

Key words

Genetic disorders are diseases like cystic fibrosis or sickle cell disease which can only be inherited and not transferred from one person to another.

Communicable diseases are diseases like athlete's foot or the common cold which can be transferred from one person to another.

A **sufferer** is a person who has inherited a genetic disorder.

A **carrier** is a person who is heterozygous for a recessive genetic disorder so they do not have the symptoms but can pass it to their children.

Benign: A slow-growing, non-cancerous tumour that cannot spread. Most benign tumours are not life-threatening.

Effects of mutation

Mutations change the genetic make-up of an organism. If this is in a body cell it is unlikely to be inherited by offspring. However, if a sperm or egg (sex cell) suffers a mutation, then this could be passed on to future generations. It would lead to increased genetic variation.

Some mutations result in **genetic disorders**. These are medical conditions that are passed from parents to offspring. They are not **communicable diseases** which can be 'caught'. We call people who have genetic disorders **'sufferers'**. Many genetic disorders are homozygous recessive. That means you need to inherit a recessive allele from both parents to have the disorder. Cystic fibrosis is an example of a genetic disorder that results from a mutation in a single gene and is homozygous recessive. A sufferer of cystic fibrosis has excess mucus in their lungs and digestive systems. We use the letter 'c' for the genotypes of cystic fibrosis.

CC means a dominant normal gene was inherited from both parents. This is called a homozygous dominant genotype. The phenotype of these alleles is normal. The person doesn't have cystic fibrosis.

cc means a recessive cystic fibrosis gene was inherited from both parents. This is called a homozygous recessive genotype. The phenotype of these alleles is a sufferer. They have cystic fibrosis.

Cc means a dominant normal gene was inherited from one parent and a recessive cystic fibrosis gene was inherited from the other parent. We call this a heterozygous genotype. Because normal is dominant over cystic fibrosis, the phenotype of these alleles is normal. This person does not have cystic fibrosis. However, this parent can pass the disorder to their children so we call them a **carrier**.

Two carriers for a genetic disorder have a 25% chance of having a child with the medical disorder. This is shown in the Punnett square below.

	C	c
C	CC	Cc
c	Cc	cc

Genotype	Description	Percentage
CC	Homozygous dominant	25%
Cc	Heterozygous	50%
cc	Homozygous recessive	25%

Phenotype	Percentage
Normal	75%
Sufferer	25%

Sickle cell disease is another genetic disorder that has arisen from a mutation. Sufferers of this disorder have red blood cells which are shaped like sickles and not the usual biconcave shape. Sickle cell disease is an especially interesting mutation because its sufferers also have an advantage. Because their red blood cells are a different shape, they are more resistant to malaria.

Sickle cell disease is inherited in the same way as cystic fibrosis above.

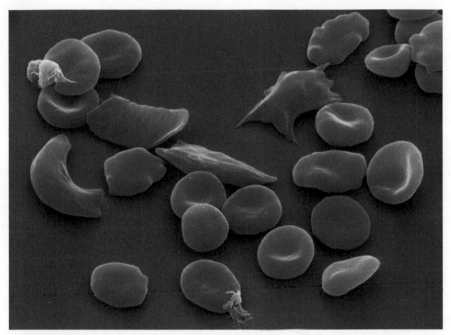

▲ Red blood cells of sufferers of sickle cell disease have red blood cells shaped like sickles

Know >

1 What is a mutation?

2 What are carcinogens?

3 What is an example of a carcinogen?

4 What do we call a person who is heterozygous for a recessive genetic disorder?

Apply >>

5 What is the difference between malignant and benign tumours?

6 What is the difference between genetic disorders and communicable diseases?

Extend >>>

7 Do some research to find out about other genetic disorders, such as Down syndrome and haemophilia.

» Extend: Cell division

I hope the answer to the question 'how old are you?' is an easy one to answer. But how about this one 'how old are your cells?'. This is a much harder question to answer because it depends upon the type of cell.

In fact, however old you are, most of your cells are about ten years old or less. Most of the tissues in our bodies are constantly being renewed. The dust that seems to relentlessly gather in your bedroom is mainly made of your dead skin cells that have dropped off after they have been replaced. Skin cells last about a month before being replaced. We make over a trillion new skin cells every month. So how do we do this?

Mitosis

Your body in made almost entirely of diploid body cells. These are all your cells except your red blood cells, which contain no DNA, and your sperm or eggs which only contain half your DNA. All other body cells have a nucleus with twenty-three pairs of chromosomes. They have your entire genome in them. The replication of these cells is called mitosis.

Both diploid daughter cells produced in mitosis are identical. You learned on the previous page that any errors here can lead to mutations.

Growth or replacement of cells

Immediately after the sperm fertilised the egg you were one diploid cell. Now you are over 30 million billion cells. All of these, except sperm or eggs, have been made by mitosis. When you are growing both of the daughter cells produced in mitosis are kept.

If you are fully grown your body still undergoes mitosis to replace your cells as they die. Now the original daughter cell dies and the new one is retained.

Meiosis

Meiosis can stand for Making Eggs In Ovaries, Sperm In Scrotum. (This isn't quite right because sperm are made in the testicles within the scrotum.)

You have one type of haploid cell; either sperm or eggs. These cells are called gametes. They fuse during fertilisation to produce a new life with a unique genome (unless you are an identical twin). Because of this, they must have half of your DNA in.

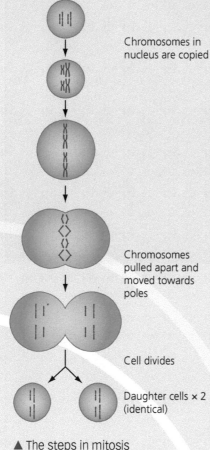

Chromosomes in nucleus are copied

Chromosomes pulled apart and moved towards poles

Cell divides

Daughter cells × 2 (identical)

▲ The steps in mitosis

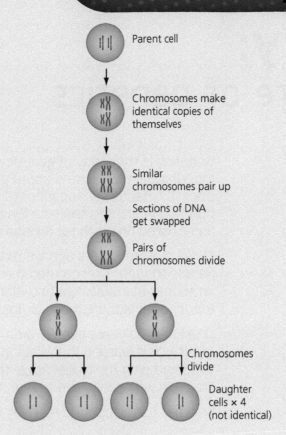

Parent cell

Chromosomes make identical copies of themselves

Similar chromosomes pair up

Sections of DNA get swapped

Pairs of chromosomes divide

Chromosomes divide

Daughter cells × 4 (not identical)

▲ The steps in meiosis

All four haploid daughter cells made in meiosis are slightly different. Unless you are an identical twin, you are not a clone of your brother or sister. This means that your parent's gametes must have been slightly different.

Amazingly, all women complete their meiosis whilst they are in their mother's womb. After this they are unable to add to the eggs in their ovaries. Men, on the other hand, are often able to produce sperm from puberty to the day they die. However, as men age the quality of their sperm decreases.

Tasks

1. What process makes identical body cells?
2. What are gametes?
3. What is true about the DNA of daughter cells of mitosis?
4. What can meiosis stand for?
5. Why does mitosis occur?
6. Why are gametes haploid?

Enquiry:
The future of genetics

Genetics is the study of genetic variation (or differences) and inheritance. You learned about genetic variation in Pupil's Book 1, Chapter 19 and inheritance earlier in this chapter. One hundred years ago seems a long time to most people. But life has existed on Earth for billions of years.

Three hundred years ago we didn't understand evolution. Many people thought God created the Universe in seven days, as it written in the Bible. Now, so many scientists believe Darwin's theory of evolution, it has almost become a fact.

One hundred years ago we didn't even understand the structure of DNA. You learned that Watson and Crick determined this in 1953 and won the Nobel Prize. They used X-ray images taken by Franklin, who sadly died before she was awarded the Nobel Prize.

So what has happened recently in genetics?

The Human Genome Project

In 2003 the Human Genome Project was completed. Hundreds of scientists from universities all over the world took thirteen years to sequence every one of over three billion DNA base pairs of randomly selected men and women in the correct order. This means that they worked out the order of every single A–T, T–A, C–G and G–C that make a human. This is an excellent example of collaborative scientific working. The results of the Human Genome Project are available to us all on the internet.

Since 2003, scientists have been following up this work by identifying where genes for specific genetic disorders are found. As a part of this they discovered that humans are only made from about 20,000 genes. This means that either large parts of our DNA don't have a purpose, or that we haven't found it yet.

Scientists have discovered the positions of genes which cause genetic disorders like cystic fibrosis and sickle cell disease. Further research hopes to develop genetic medicines that are specific to individual people. Imagine if instead of being given the same medicine as everyone else, you got a specialised medicine based upon your own DNA. It might not sound possible, but neither did evolution or the structure of DNA before they were discovered.

Genetic modification

Key word

Genetic modification is when a gene for a desirable characteristic is inserted from one species into another.

Genetic modification is the removal of a gene or genes from one species and insertion into the genome of another organism. This was first completed in bacteria in 1973 by American scientists Herbert Boyer (1936–) and Stanley Cohen (1935–). In the same year the first animal (a mouse) was genetically modified. A decade later tobacco became the first plant to be genetically modified.

A famous example of genetic modification involves the glow-in-the-dark gene from a jellyfish. This has been removed from a jellyfish and inserted into the DNA of mouse and rabbit embryos. The embryos are genetically modified because every cell that then grows from it will have the modification. It would be almost impossible to genetically modify all cells of adult organisms.

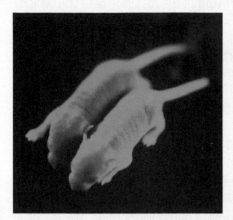

▲ Genetic modified mice have the glow-in-the-dark gene from jellyfish

Glow-in-the-dark animals might sound interesting but what use do they have? The glow-in-the-dark gene is often added alongside a more useful gene. Why do you think this is? Because if the adult organism glows in the dark scientists know the other gene has been inserted as well. Useful examples of genetic modification include the following.

- Inserting the human gene for insulin into bacteria that now produce insulin for diabetics.

- Golden rice now contains carotene from carrots which helps people make vitamin A.

- Soya beans have been genetically modified to be resistant to a herbicide weedkiller so farmers can quickly spray their crops and kill only weeds.

Genetic modification is an ethical issue. This is one that some people disagree with for religious or moral reasons. It is illegal to genetically modify humans.

One hundred years ago we couldn't have predicted that we would discovered the structure of DNA. Three hundred years ago we could say the same about evolution. Who know what we will discover in the next lifetime. This is exactly why science is so interesting and so very important.

1. Why are some people against genetic modification?
2. What is your genetic code made from?
3. What is a genome?
4. What is a gene?
5. What is genetic modification?
6. How have rabbits and mice been genetically modified?
7. Apart from mapping the human genome, what was another major success of the Human Genome Project?
8. What genetic modification would you make to your pet?
9. What do you think of the Human Genome Project?
10. What do you think of genetic modification?
11. What other examples of genetic modification can you research?

Glossary

A

Absorption is when some or all of the energy of a wave is transferred to a material.

Aerobic respiration is the process of breaking down glucose with oxygen to release energy and produces carbon dioxide and water.

Alleles are alternative forms of the same gene.

Alveoli are the small air sacs found at the end of each bronchiole.

Amplitude is the height of a wave, measured from the middle.

Anaerobic respiration (fermentation) is the process of releasing energy from the breakdown of glucose without oxygen, producing lactic acid (in animals) and ethanol and carbon dioxide (in plants and microorganisms).

Atmospheric pressure is the pressure caused by the weight of the air above a surface.

An **atom** is the smallest possible part of an element.

An **attractive force** pulls things together.

B

Bias is when a researcher controls the outcome of an investigation, or when a journal favours a particular point of view.

Biodiversity is the variety of living things. It is measured as the differences between individuals of the same species, or the number of different species in an ecosystem.

Breathing is the movement of air in and out of the lungs.

The **bronchi** are two tubes which carry air to the lungs.

Bronchioles are small tubes in the lung.

C

Carbohydrates are the body's main source of energy. There are two types: simple (sugars) and complex (starch).

A **carbon sink** is an area of vegetation, the ocean or the soil, which can absorb and store carbon.

A **carrier** is a person who is heterozygous for a recessive genetic disorder so they do not have the symptoms but can pass it to their children.

A **catalyst** is a chemical which speeds up a reaction but is not used up or chemically changed.

Chemical bonds are strong forces that hold two atoms together.

The **chemical formula** of a substance tells you which elements are present, and how many atoms of each there are.

Chemical properties are a description of the way that a substance reacts with other chemicals. Being reactive is a chemical property. So is being acidic.

In **chemical reactions**, a new substance is made because chemical bonds are broken and made.

Chlorophyll is the green pigment in plants and algae that absorbs light energy for photosynthesis

Chromosomes are thread-like structures containing tightly coiled DNA.

Combustion is another name for burning. This is when a fuel reacts with oxygen and releases heat and light energy.

Competition is when two or more living things struggle against each other to get the same resource.

A **compound** is a pure substance made from two or more elements which are chemically combined in a fixed ratio of atoms.

Compression is the name given to forces that squash or push together.

Communicable diseases are diseases like athlete's foot or the common cold which can be transferred from one person to another.

Conduction is the transfer of energy by the vibration and collision of particles.

Mass is **conserved** in a chemical reaction because it stays the same. The total mass of atoms before and after the reaction is the same.

A **contact** force is one which acts only when objects are in contact with each other. Once the contact is removed the force is no longer applied.

Convection is the transfer of energy when the particles in a heated fluid rise.

The **core** of an electromagnet is a piece of soft iron which the solenoid is wrapped around.

A **correlation** is a relationship between two variables. The dependent variable increases or decreases as the independent variable increases.

The flow or movement of charge is called **current** and is measured in amperes (A).

D

Deformation occurs when an object's shape is changed due to a force. An elastic object can be stretched or squashed by an applied force. Work is done to deform the object.

The **dependent variable** is the factor you measure in an investigation.

The **diaphragm** is a sheet of muscle found underneath the lungs.

Dietary fibre is the parts of plants that cannot be digested, which helps the body eliminate waste.

All cells except sperm and eggs are **diploid** cells which contain all an organism's DNA (its genome).

There is a linear relationship between two variables if there is a straight line when they are plotted on a graph. If the straight line goes through the origin, then they are described as **directly proportional** to each other.

Displacement is the distance an object moves from its starting point as measured in a straight line from that point.

DNA is a molecule found in the nucleus of cells that contains genetic information.

A **dominant** characteristic is one that is observed even if only one allele is present.

Drag occurs when two surfaces move over each other and one of the surfaces is a fluid. It is a frictional force which opposes motion.

E

Electrolysis is when a compound is broken down into elements using electricity.

An **electromagnet** is a non-permanent magnet turned on and off by controlling the current through it.

Electromagnetism is the temporary magnetic field caused by an electrical current in any material.

An **element** is a pure substance made from only one type of atom. Elements are listed on the periodic table.

In an **endothermic reaction**, energy is taken in, usually as heat.

Enzymes are proteins made by living organisms to speed up chemical reactions that take place within cells.

An object is in **equilibrium** if the opposing forces acting on it are balanced.

An **estimate** is when you make an educated guess at a value, using scientific ideas, data or a rough calculation to help you to decide on your chosen value.

Evidence is an observation or measurement which supports a hypothesis.

Evolution is the theory that the animal and plant species living today descended from species that existed in the past.

In an **exothermic reaction**, energy is given out, usually as heat or light.

To **extract** a metal is to separate it from a compound or mixture.

Extinct is when no more individuals of a species remain.

F

Fertilisers are chemicals containing minerals that plants need to build new tissues.

A **finite resource** is a natural resource that might run out in the future if humans are not careful about how quickly it is used up.

A **fluid** is a substance that has no fixed shape. Both gases and liquids are fluids.

Fossil fuels are non-renewable energy resources formed from the remains of ancient plants or animals. Examples are coal, crude oil and natural gas. When a fossil fuel is burned, carbon dioxide is released into the atmosphere.

The number of complete waves detected in one second is called the **frequency**.

Friction is the force between two surfaces that are sliding over each other, or that are trying to slide over each other. Friction always opposes the direction of motion.

A **fuel** is a substance that can be burned to release energy.

A **funder** is a person or organisation which pays for scientific research to be done.

G

A **gene** is a section of DNA that determines an inherited characteristic.

Genetic disorders are diseases like cystic fibrosis or sickle cell disease which can only be inherited and not transferred from one person to another.

Genetic modification is when a gene for a desirable characteristic is inserted from one species into another.

The **genome** is one copy of all of an organism's DNA, found in every diploid body cell.

The **genotype** is the genetic makeup of an organism, given as a pair of letters to denote alleles of a gene.

Global warming is the gradual increase in surface temperature of the Earth.

The **greenhouse effect** is when energy from the Sun is transferred into the thermal energy store of gases in the atmosphere.

A **group** on the periodic table is a vertical column.

Gut bacteria are microorganisms that naturally live in the intestine and help food break down.

H

Sperm and eggs in animals (and pollen and eggs in plants) are **haploid** cells which contain half an organism's DNA.

Heterozygous is the presence of one dominant and one recessive allele.

Homozygous dominant is the presence of two dominant alleles.

Homozygous recessive is the presence of two recessive alleles.

A **hypothesis** is a prediction which can be tested by experiments or observations.

I

An **incompressible liquid** is an ideal liquid which doesn't change volume when a force is applied. Most real liquids, such as water, will be squashed slightly if a force is applied.

The **independent variable** is the factor you change in an investigation.

A chemical is **inert** if it doesn't react with other chemicals.

Infrared (IR) waves have frequencies lower than visible light.

Infrasound waves are outside of the human auditory range, by having very low frequencies.

Inherited characteristics are features that are passed from parents to their offspring.

The **input force** is the force you apply to the machine.

The **interval** is the gap between values of a variable.

K

A **kinetic store** of energy is filled when an object speeds up. This is also known as kinetic energy.

L

The **large intestine** is the lower part of the intestine from which water is absorbed and where faeces are formed.

A **lever** is a simple mechanism that can be used to allow a small force to have a larger effect. It pivots about a point, known as a fulcrum.

There is a **linear relationship** between two variables if there is a straight line when they are plotted on a graph. If the straight line goes through the origin, then they are described as directly proportional to each other.

Lipids (fats and oils) are a source of energy and are found in butter, milk, eggs and nuts.

The displacement of a **longitudinal wave** is along or in line with the direction of wave travel.

A **loudspeaker** converts an electrical signal into a sound wave.

The **lung volume** is a measure of the amount of air breathed in or out.

M

A **microphone** converts a sound wave into an electrical signal.

A **mineral** is a solid chemical compound found in the Earth's crust. Because each mineral is a compound, it will have a specific chemical formula.

A **molecule** is a tiny particle of a compound. Molecules are made from atoms which are strongly bonded together.

Mutation is changing of the structure of a gene.

N

Natural resources are substances from the Earth which act as raw materials for making a variety of products.

Natural selection is the process by which species change over time in response to environmental changes and competition for resources.

Non-destructive testing is used when the object being tested must not be damaged by the test.

O

An **ore** is a rock which contains enough of a mineral to make it worth extracting from the crust.

The **output force** is the force that is applied to the object being moved by the machine.

P

A journal is a magazine which publishes science research. Almost all journals are **peer reviewed**, which means the articles are checked by other expert scientists before being printed.

A **period** on the periodic table is a horizontal row.

The **periodic table** lists all of the chemical elements in groups and periods.

A **permanent magnet** is magnetic all the time.

The **phenotype** is the physical characteristics of an organism's genotype.

Photosynthesis is the process that uses the Sun's energy to convert carbon dioxide and water into glucose and oxygen.

In **physical changes**, the properties of a substance change but no new substance is made. Changes of state and dissolving are all examples of physical changes.

Physical properties are a description of the way that a substance behaves which don't involve it reacting with other chemicals. Melting point, electrical conductivity and hardness are examples of physical properties.

A **polymer** is a very long molecule made from thousands of smaller molecules joined together in a repeating pattern.

A **population** is a group of the same species living in an area.

Pressure is the ratio of force to surface area for a fluid. The unit is N/m^2 or Pascals (Pa).

A **pressure wave** like sound has repeating patterns of high-pressure and low-pressure regions.

Products are substances which are formed in a reaction. They are always after the arrow in an equation.

Proteins are nutrients your body uses to build new tissue for growth and repair. Sources are meat, fish, eggs, dairy products, beans, nuts and seeds.

R

Radiation is a method of transferring energy as a wave.

Reactants are substances that react together in a reaction. They are always before the arrow in an equation.

A **recessive** characteristic is one that is only observed if no dominant alleles are present.

Recycling is when an object is processed so that the materials it is made from can be used again.

Reflection is when a wave changes direction away from a surface.

Refraction is a change in the direction of light when it goes from one material into another. The new direction, measured from the normal, is the angle of refraction.

A **repulsive force** pushes things apart.

Resistance measures how hard it is to push charges through a material or component. It is measured in ohms (Ω).

The **respiratory system** replaces oxygen and removes carbon dioxide from blood.

A **resultant force** is a single force that could replace all of the forces acting on an object and still have the same effect on the object.

The **ribs** are bones that surround the lungs to form the ribcage.

S

Secondary data are results which were collected by someone else. **Secondary research** may also relate to conclusions made by someone else.

The **small intestine** is the upper part of the intestine where digestion is completed and nutrients are absorbed by the blood.

A **solenoid** is a wire wound into a tight coil, part of an electromagnet.

A **species** is a group of organisms that can interbreed to have fertile offspring.

The **stomach** is a sac where food is mixed with acidic juices to start the digestion of protein and kill microorganisms.

Stomata are pores in the bottom of a leaf which open and close to let gases in and out.

The **stress** on a surface is the ratio of applied force to surface area. The units of stress are N/m^2.

A **sufferer** is a person who has inherited a genetic disorder.

T

The **temperature** of an object is a measure of the motion and energy of the particles that form the object.

Tension is the name given to forces that extend or pull apart.

A **thermal conductor** is a material that increases how fast energy is transferred between objects.

Thermal decomposition is when a single reactant breaks down into simpler substances when it is heated.

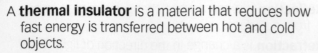

A **thermal insulator** is a material that reduces how fast energy is transferred between hot and cold objects.

A **thermal store of energy** is a measure of the energy stored in a substance due to the vibration and motion of particles. It is sometimes just called thermal energy.

Thermochromic film changes colour depending on the temperature. It can be used as a simple thermometer.

The **trachea** (windpipe) carries air from the mouth and nose to the lungs.

Transmission is when the wave goes through the material instead of being absorbed or reflected.

Transverse waves like light and ripples on water have displacement which is perpendicular or at a right angle to the direction of wave movement.

U

Ultrasound waves have frequencies higher than the human auditory range.

Ultraviolet (UV) waves have frequencies higher than visible light.

Upthrust is the upward force that a liquid or gas exerts on an object floating in it.

V

Variation is the differences within and between species.

W

A **wave** always involves the transfer of energy by vibrations from one place or material to another, without matter being permanently moved.

The **wavelength** is the length of one complete wave pattern. It is measured in metres (m) and has the symbol λ.

Work is the transfer of energy when a force moves an object in the direction of the force. Work is measured in joules.

Index

V

variation 228, 232, 233
Visking tubing 197
vitamins 187, 213

W

water 113
    adaptations to aquatic environments 227
    alkali metals' reactions with 105
    needed for life 200
    needed by plants for growth 213
    photosynthesis 149, 214, 215
water cycle 148
water lilies 227
water waves 86–7
wave equation 94–5
wave power 87
wavelength 83, 86, 90, 94–5
waves
    effects 79, 80–7
    properties 79, 88–97
wheels 57
white rhinos 237

work 53, 54–63
    calculating 60–1

X

xylem vessels 218–19

Y

yeast 206–7, 210–11

Photo credits

The Publisher would like to thank the following for permission to reproduce copyright material:

p.6 © Africa Studio / Shutterstock.com; **p.8** © Imagestate Media (John Foxx) Leisure Time V3072; **p.9** © DGB / Alamy Stock Photo; **p.10** © C. Davenport; **p.12** *tl* © jonnysek / Fotolia, *tr* © Lsantilli / Fotolia, *bl* © Focus Pocus LTD / Fotolia, *br* © MARK CLARKE / SCIENCE PHOTO LIBRARY; **p.15** © joegast / Fotolia; **p.16** © dpa picture alliance archive/Alamy Stock Photo; **p.17** *t* © C. Davenport, *m* © J. J. Sanderson / Northumbria University; **p.18** © NASA; **p.19** © C. Davenport; **p.20** *l* © GIPhotoStock / SCIENCE PHOTO LIBRARY, *r* © SCIENCE PHOTO LIBRARY; **p.21** *t* © swisshippo / Fotolia, *b* © C. Davenport; **p.22** © Gina Sanders / Fotolia; **p.23** © Africa Studio / Shutterstock.com; **p.24** © MARTYN F. CHILLMAID / SCIENCE PHOTO LIBRARY; **p.25** *l* © S.Pytel / Shutterstock.com, *r* © GIPHOTOSTOCK / SCIENCE PHOTO LIBRARY; **p.26** © Martyn F Chillmaid; **p.27** *l* © CHARLES D. WINTERS / SCIENCE PHOTO LIBRARY, *r* © BRITISH ANTARCTIC SURVEY / SCIENCE PHOTO LIBRARY; **p.28** © MHjerpe - Fotolia.com; **p.29** *m* © DANTE FENOLIO / SCIENCE PHOTO LIBRARY, *b* © USS TUNNY (SSN 682) http://www.ssn682.com; **p.30** © Dpa picture alliance / Alamy Stock Photo; **p.31** © C. Davenport; ; **p.32** © Phil Degginger / SCIENCE PHOTO LIBRARY; **p.34** © MilanB / Shutterstock.com; **p.36** *l* © Phil Degginger / SCIENCE PHOTO LIBRARY, *r* © Rick Parsons; **p.37** *l* © Andreas Gradin – Fotolia, *r* © Tatiana Buzuleac - iStock via Thinkstock; **p.38** © MilanB / Shutterstock.com; **p.42** © GIPHOTOSTOCK / SCIENCE PHOTO LIBRARY; **p.45** © GIPHOTOSTOCK / SCIENCE PHOTO LIBRARY; **p.46** © TFoxFoto / Shutterstock; **p.50** *t* © iofoto / Fotolia, *m* © Dorling Kindersley / UIG / SCIENCE PHOTO LIBRARY; **p.51** © PPP / Fotolia; **p.52** ©Iakov Kalinin - Fotolia; **p.54** © BlueOrange Studio / Shutterstock.com; **p.55** © Podushko Alexander / Shutterstock.com; **p.56** *l* © NASA images, *r* © Neil Dixon; **p.57** *t* © kris Mercer / Alamy Stock Photo, *m* © Christine Whitehead / Alamy Stock Photo, *b* © Dmitro / Shutterstock.com; **p.60** *m* © dpa picture alliance / Alamy Stock Photo, *b* © WoodysPhotos / Shutterstock.com; **p.61** *t* © Ivan Smuk / Shutterstock.com, *ml* © Manutsawee Buapet / Shutterstock.com, *mr* © Patxi – Fotolia.com; **p.62** *tl* © Ilona Ignatova / Shutterstock.com, *tr* © Leonid Andronov – Fotolia; **p.63** © kangshutters / Shutterstock; **p.64** *l* © Glenn Bo / iStockphoto / Thinkstock, *r* © Morchella / Shutterstock.com; **p.66** © Fotolia; **p.67** © lee hacker / Alamy Stock Photo; **p.68** © SCIENCE PHOTO LIBRARY; **p.69** © SCIENCE PHOTO LIBRARY; **p.70** *l* © marytmoore / Fotolia.com, *r* © veneratio – Fotolia.com; **p.71** © MARTYN F. CHILLMAID / SCIENCE PHOTO LIBRARY; **p.72** © Korionov / Shutterstock.com; **p.73** *t* © Iakov Kalinin – Fotolia, *b* ©AZP Worldwide – Fotolia; **p.74** *t* © MARK SYKES / SCIENCE PHOTO LIBRARY, *ml* © EDWARD KINSMAN/SCIENCE PHOTO LIBRARY, *mr* © Roman Ivaschenko/stock.adobe.com; **p.75** *t* © PHILIPPE PSAILA / SCIENCE PHOTO LIBRARY, *b* © Albert Nowicki / Shutterstock; **p.76** *l* © VICTOR DE SCHWANBERG / SCIENCE PHOTO LIBRARY, *r* © C. Davenport; **p.77** *l* © dvoevnore / Shutterstock.com, *r* © moodboard / Thinkstock / Getty Images; **p.78** © rebelml / iStock / Thinkstock / Getty Images; **p.80** *tl* © GEORGY SHAFEEV / SCIENCE PHOTO LIBRARY, *tr* © Ian Allenden / 123rf, *ml* ©roblan – Fotolia, *mr* © ZEPHYR / SCIENCE PHOTO LIBRARY; **p.81** *t* © yoshiyayo / 123RF, *b* ©dav820 – Fotolia.com; **p.82** ©TEA - Fotolia; **p.84** © H. S. Photos / SCIENCE PHOTO LIBRARY; **p.85** ©Photographer: Shane White; **p.86** *t* © rebelml / iStock / Thinkstock / Getty Images, *b* © Imagestate Media (John Foxx) / Splash V3060; **p.88** *t* ©f9photos – Fotolia, *b* © carol.anne / Shutterstock; **p.89** © ANDREW LAMBERT PHOTOGRAPHY / SCIENCE PHOTO LIBRARY; **p.92** *tl* © claudiodivizia – iStock – Thinkstock, *tr* © jcg_oida – Fotolia, *m* ©ARochau – Fotolia; **p.94** © Gelly Images / Image Source – OurThreatened Environment IS236; **p.95** © Marques / Shutterstock.com; **p.96** *t* ©Daniel Schweinert / Fotolia, *m* © ASHLEY COOPER / SCIENCE PHOTO LIBRARY, *b* ©Watchtheworld / Alamy Stock Photo; **p.97** © Andrey Ezhov / Shutterstock.com; **p.98** © Nick Dixon; **p.103** *l* © mountain beetle / Shutterstock.com, *r* © Africa Studio / Shutterstock.com; **p.105** © Neil Dixon; **p.106** *all* © Neil Dixon; **p.109** © ANDREW LAMBERT PHOTOGRAPHY / SCIENCE PHOTO LIBRARY; **p.110** *m&b* © SCIENCE PHOTO LIBRARY; **p.111** *l* © SPUTNIK / SCIENCE PHOTO LIBRARY; **p.112** © Nick Dixon; **p.113** © Nick Dixon **p.118** *t* © ILEISH ANNA / Shutterstock.com, *b* © Hugh Peterswald / Alamy Stock Photo; **p.119** *t* © Vadim Ratnikov / Shutterstick.com, *m* © Crevis / Shutterstock.com, *b* © Max Earey / Shutterstock; **p.120** *t* © PHILIPPE PLAILLY / SCIENCE PHOTO LIBRARY, *b* © Neil Dixon; **p.121** © Neil Dixon; **p.122** © Neil Dixon; **p.124** *l* © Geoffrey Whiteway / 123RF, *r* © Neil Dixon; **p.125** © Neil Dixon; **p.126** © Neil Dixon; **p.132** © Neil Dixon; **p.133** *t&m* © Neil Dixon; **p.134** © Zadiraka Evgenii / Shutterstock.com; **p.135** © Neil Dixon; **p.136** © © David Franklin / Shutterstock.com; **p.138** © CAIA IMAGE / SCIENCE PHOTO LIBRARY; **p.142** *l&r* © Neil Dixon; **p.143** © Neil Dixon; **p.144** *l&r* © Neil Dixon; **p.146** © Natapong Ratanavi / Shutterstock.com; **p.151** © stocker1970 / Shutterstock.com; **p.152** © muratart / Shutterstock.com; **p.154** © Natapong Ratanavi / Shutterstock.com; **p.155** © Microgen / Shutterstock.com; **p.156** © Glynnis Jones / Shutterstock.com; **p.157** © Wikipedia™; **p.159** © Neil Dixon; **p.161** © cozyta / Shuttwestock.com; **p.162** *ml&mr* © Neil Dixon, *b* © GIPhotoStock / SCIENCE PHOTO LIBRARY; **p.163** © PETR JAN JURACKA / SCIENCE PHOTO LIBRARY; **p.164** *t* © SCIENCE PHOTO LIBRARY, *m* © Neil Dixon, *b* © PAUL D STEWART / SCIENCE PHOTO LIBRARY; **p.166** *t* © GraphicsRF / Shutterstock.com, *bl* © 8winnond / Shutterstock.com, *br* © Neil Dixon; **p.167** *t* © Marius GODOI / Shutterstock.com, *m* © Evan Lorne / Shutterstock.com, *b* © posteriori / Shutterstock.com; **p.168** *t* © Federico Rostagno / Shutterstock.com, *b* © JUAN GAERTNER / SCIENCE PHOTO LIBRARY; **p.169** *l* © ASHLEY COOPER / SCIENCE PHOTO LIBRARY, *r* © Becky Stares / Shutterstock.com; **p.170** © NIBSC / SCIENCE PHOTO LIBRARY; **p.174** *m* © DAVID M. MARTIN, MD / SCIENCE PHOTO LIBRARY, *b* © NIBSC / SCIENCE PHOTO LIBRARY; **p.176** © DAVID MCCARTHY / SCIENCE PHOTO LIBRARY; **p.180** *t* © DAVID MCCARTHY / SCIENCE PHOTO LIBRARY, *m* © Nick Dixon, *b* © Nick Dixon; **p.181** *l&r* © MARTYN F. CHILLMAID / SCIENCE PHOTO LIBRARY; **p.183** © CROWN COPYRIGHT / HEALTH & SAFETY LABORATORY / SCIENCE PHOTO LIBRARY; **p.184** *m* © ggw – Fotolia.com, *b* © Maridav / iStock / Thinkstock / Getty Images; **p.185** *t* © creadorimatges – Fotolia, *m* © Phanie / Alamy Stock Photo; **p.186** *l* © Karen Struthers – Fotolia, *r* © Africa Studio – Fotolia; **p.187** © Robyn Mackenzie / 123RF; **p.188** © ifong / Shutterstock.com; **p.189** © Christos Georghiou / Sahutterstock.com; **p.190** *t* © GASTROLAB / SCIENCE PHOTO LIBRARY, *m* © DAVID M. MARTIN, MD / SCIENCE PHOTO LIBRARY; **p.191** © GASTROLAB / SCIENCE PHOTO LIBRARY; **p.193** *l* ©DENNIS KUNKEL MICROSCOPY / SCIENCE PHOTO LIBRARY, *r* © GARO / PHANIE / SCIENCE PHOTO LIBRARY; **p.196** *t* © JohnnyGreig / E+ / Getty, *m* © Neil Dixon, *b* © Nick Dixon; **p.197** ©TREVOR CLIFFORD PHOTOGRAPHY / SCIENCE PHOTO LIBRARY; **p.198** © John Anderson – Fotolia; **p.200** *m* © PaulPaladin – Fotolia, *b* © kichatof – Fotolia.com; **p.201** © Lakeview Images / Shutterstock; **p.202** © Photographer:Stephane Bidouze; **p.203** *t* © Dinadesign – Fotolia, *m* © DENNIS KUNKEL MICROSCOPY / SCIENCE PHOTO LIBRARY; **p.204** © Fotolia; **p.205** © John Fryer / Alamy

Stock Photo; **p.206** *t* © Guy Erwood – Fotolia, *b* © bwf211 / 123RF.com; **p.207** © adisa – Fotolia; **p.209** *t&b* © Getty Images / Thinkstock / iStockphoto / ZambeziShark; **p.210** © hjschneider – Fotolia; **p.211** © SCIENCE PHOTO LIBRARY; **p.212** © Getty Images / Goodshoot RF / Thinkstock; **p.213** *t* © John Anderson – Fotolia, *m* © kaetana / Shutterstock.com; **p.214** © Diana Taliun / Shutterstock.com; **p.215** *t* © Diana Taliun / Shutterstock.com; *m* © LEX – Fotolia; **p.216** © TairA / Shutterstock.com; **p.217** © D. Kucharski K. Kucharska / Shutterstock.com; **p.218** *m* © POWER AND SYRED / SCIENCE PHOTO LIBRARY, *b* © Filip Fuxa / Shutterstock.com; **p.221** © P RONA / OAR / NATIONAL UNDERSEA RESEARCH PROGRAM / NOAA / SCIENCE PHOTO LIBRARY; **p.222** © © Yio – Fotolia; **p.223** © CORDELIA MOLLOY / SCIENCE PHOTO LIBRARY; **p.224** © Ronnie Howard – Fotolia; **p.226** *t* © Getty Images / Thinkstock / iStockphoto / ZambeziShark, *m* © paolofusacchia – Fotolia, *b* © Getty Images / Thinkstock / iStockphoto / ZambeziShark; **p.227** *l* © Getty Images / Goodshoot RF / Thinkstock, *m* © Vera Kuttelvaserova Stuchelova / 123RF, *r* © Imagestate Media (John Foxx) / Wonderful Flowers SS89; **p.228** © RoyA – Fotolia; **p.229** © Hogan Imaging – Fotolia; **p.230** © Georgios Kollidas – Fotolia; **p.231** *t* © Classic Image / Alamy Stock Photo, *m* © FineArt / Alamy Stock Photo, *b* © Sipa Press / REX / Shutterstock; **p.232** *tl* © alain – Fotolia, *tm* © MR1805 – iStock via Thinkstock / Getty Images, *tr* © Nico Smit / Fotolia.com, *bl* © Ronnie Howard – Fotolia, *bm* © paolofusacchia – Fotolia, *br* © ReformBoehm – iStock – Thinkstock / Getty Images; **p.233** *t* © Karina Baumgart – Fotolia; *m* © Dalshe / Shuttwerstock.com; **p.234** © MICHAEL W. TWEEDIE / SCIENCE PHOTO LIBRARY; **p.235** © MICHAEL W. TWEEDIE / SCIENCE PHOTO LIBRARY; **p.236** © Mopic / Shutterstock.com; **p.237** *m* © Jamie Ahmad / Shutterstock.com, *b* © The Washington Post/Nichole Sobecki/Getty Images; **p.238** *m* © Marco Maggesi / Shutterstock.com, *b* © Paul Nash / Shutterstock.com; **p.239** *l* © Shelly Still / Shutterstock.com, *m* © Patrick K. Campbell / Shutterstock.com, *r* © Matt Jeppson / Shutterstock.com; **p.240** © Luk Cox – Fotolia.com; **p.243** © m.arc – Fotolia, **p.244** *t* © Mateusz Kopyt / Shutterstock.com, © *b* © Luk Cox – Fotolia.com; **p.245** © TheSupe87 – Fotolia; **p.246** © Elflaco1983 / Shutterstock.com; **p.247** © EYE OF SCIENCE / SCIENCE PHOTO LIBRARY; **p.251** © MAKOTO IWAFUJI / EURELIOS / SCIENCE PHOTO LIBRARY.

t = top, *b* = bottom, *l* = left, *r* = right, *m* = middle

Every effort has been made to trace all copyright holders, but if any have been inadvertently overlooked, the Publisher will be pleased to make the necessary arrangments at the first opportunity.